Image Segmentation for Environmental Monitoring

Image Segmentation for Environmental Monitoring

Special Issue Editors

Brian Alan Johnson
Lei Ma

MDPI • Basel • Beijing • Wuhan • Barcelona • Belgrade • Manchester • Tokyo • Cluj • Tianjin

Special Issue Editors

Brian Alan Johnson
Institute for Global
Environmental Strategies (IGES)
Japan

Lei Ma
Nanjing University
China

Editorial Office
MDPI
St. Alban-Anlage 66
4052 Basel, Switzerland

This is a reprint of articles from the Special Issue published online in the open access journal *Remote Sensing* (ISSN 2072-4292) (available at: https://www.mdpi.com/journal/remotesensing/special_issues/image-segentation).

For citation purposes, cite each article independently as indicated on the article page online and as indicated below:

LastName, A.A.; LastName, B.B.; LastName, C.C. Article Title. *Journal Name* **Year**, *Article Number, Page Range.*

ISBN 978-3-03936-477-0 (Hbk)
ISBN 978-3-03936-478-7 (PDF)

Contents

About the Special Issue Editors

Brian Johnson received a M.S. in Geography from Florida Atlantic University, USA (2007), and a Ph.D. in Geosciences from Florida Atlantic University, USA (2012). He has been working at IGES since 2013, and his research involves developing and applying geographic information systems (GIS), remote sensing, and spatial modeling methods/techniques to support environmental management and land use planning efforts. He is also an Associate Editor for the journal *Remote Sensing*.

Lei Ma received a M.S. in Geographic Information Systems from Southwest Jiaotong University, Chengdu, China, in 2011, and a Ph.D. in Geography from Nanjing University, Nanjing, China, in 2016. From 2014 to 2015, he was a guest Ph.D. student with University of Salzburg, Austria. From 2016 to 2019, he was a Research Associate with Nanjing University. He was also a Post-Doctoral Research Fellow with Texas Tech University from 2018 to 2019, Lubbock, Texas, USA. Since 2019, he has been an Alexander von Humboldt Research Fellow with Technical University of Munich, and an Associate Professor with Nanjing University. His research interests include object-based image analysis, land cover and land use change, high-resolution image analysis, and time series analysis. Dr. Lei Ma received The Jack Dangermond Award–Best Paper 2017 in 2019. He was selected to receive the Jiangsu province-level outstanding diploma thesis award in 2017. He is a Guest Editor for *Remote Sensing*.

Preface to "Image Segmentation for Environmental Monitoring"

Image segmentation, as a fundamental component of object-based image analysis (OBIA), has become a major topic of interest in the environmental remote sensing field due to the ever-increasing quantity of high spatial resolution (HSR) imagery acquired from satellites, airplanes, unmanned aerial vehicles (UAVs), and other platforms. This Special Issue provides new ideas related to image segmentation methods, strategies, and applications. In addition to new segmentation methods and segmentation parameter optimization strategies, several thematic mapping studies for ecotope, urban green cover, landslide, and arid-land vegetation cover mapping are included. Finally, readers will find that this book is trying to help bridge the current gaps between segmentation methods and environmental monitoring applications.

Brian Alan Johnson, Lei Ma
Special Issue Editors

remote sensing

Editorial

Image Segmentation and Object-Based Image Analysis for Environmental Monitoring: Recent Areas of Interest, Researchers' Views on the Future Priorities

Brian Alan Johnson [1] and **Lei Ma** [2,3,*]

[1]   Natural Resources and Ecosystem Services, Institute for Global Environmental Strategies, 2108-11, Kamiyamaguchi, Hayama, Kanagawa 240-0115, Japan; johnson@iges.or.jp
[2]   School of Geography and Ocean Science, Nanjing University, Nanjing 210023, China
[3]   Signal Processing in Earth Observation, Technical University of Munich, 80333 Munich, Germany
*   Correspondence: maleinju@nju.edu.cn

Received: 25 May 2020; Accepted: 29 May 2020; Published: 1 June 2020

check for
updates

Abstract: Image segmentation and geographic object-based image analysis (GEOBIA) were proposed around the turn of the century as a means to analyze high-spatial-resolution remote sensing images. Since then, object-based approaches have been used to analyze a wide range of images for numerous applications. In this Editorial, we present some highlights of image segmentation and GEOBIA research from the last two years (2018–2019), including a Special Issue published in the journal *Remote Sensing*. As a final contribution of this special issue, we have shared the views of 45 other researchers (corresponding authors of published papers on GEOBIA in 2018–2019) on the current state and future priorities of this field, gathered through an online survey. Most researchers surveyed acknowledged that image segmentation/GEOBIA approaches have achieved a high level of maturity, although the need for more free user-friendly software and tools, further automation, better integration with new machine-learning approaches (including deep learning), and more suitable accuracy assessment methods was frequently pointed out.

Keywords: GEOBIA; object-based image analysis; high-spatial-resolution; image segmentation parameter optimization

---

1. Introduction

Image segmentation and (geographic) object-based image analysis (GEOBIA [1], or simply OBIA), have been utilized in remote sensing for around two decades now [2]. Image segmentation is the first step of GEOBIA, and involves the partitioning of an image into relatively homogeneous regions, i.e., "image segments" or "image objects" [3]. These image segments serve as the base unit for further analysis, e.g., image classification or change detection, using the spectral/spatial/contextual attributes of the segments. Image segmentation is a fundamental issue in GEOBIA research, as the quality of segmentation results often affects the accuracy of subsequent analysis (e.g., land-use/land-cover classification accuracy).

Originally, GEOBIA was proposed as a way to incorporate contextual information for high-spatial-resolution image classification, which was necessary because the pixels in these images tend to be smaller than the real-world features intended to be mapped [2,3]. Since then, it has been used to analyze images having a wide range of spatial resolutions and from various types of sensors (e.g., multispectral, hyperspectral, synthetic aperture radar). The first major review of this topic was conducted in 2010 [4], and since then several others have been undertaken [5–7].

In this Editorial, we share some highlights of GEOBIA research over the last two years (2018–2019), including a Special Issue on the topic in the journal *Remote Sensing*. We also present 45 researchers' responses to an online questionnaire on the current state and future priorities of GEOBIA research.

2. Highlights from 2018–2019

2.1. Research Topics of Interest

From a search of the Scopus database (title/keyword/abstract search for papers containing the term "object-based image analysis"), we identified 369 journal articles published on the topic of GEOBIA over the last two years (2018–2019). From these articles, we attempted to highlight some topics of significant recent interest based on the text in the papers' titles/keywords/abstracts. High-frequency terms from the text were identified using Citespace software [8], and after filtering out several overly general terms (e.g., "object", "based", "image", "analysis", "remote sensing", and "resolution"), a wordcloud map (Figure 1) was generated to allow for a visualization of the frequently-used terms (larger words in the figure were more frequently used). In Figure 1, mapping and segmentation can be seen as the most frequent areas of interest overall, which is perhaps unsurprising. The types of applications GEOBIA was most frequently used to support can be seen as forestry, vegetation, wetland, and urban area analysis. Classification algorithms that were of significant interest included decision trees (which are often incorporated in ensemble algorithms like random forests [9]) as well as support vector machines [10]. Finally, the most frequent remote sensing datasets of interest included Landsat images, synthetic aperture radar (SAR) data, Worldview images, Sentinel images, images from UAVs/other airborne optical sensors, and Lidar data. This frequent interest in moderate spatial resolution imagery (e.g., Landsat and Sentinel) as well as SAR/Lidar data suggests that GEOBIA has moved beyond its initial sole focus on high-spatial-resolution optical data.

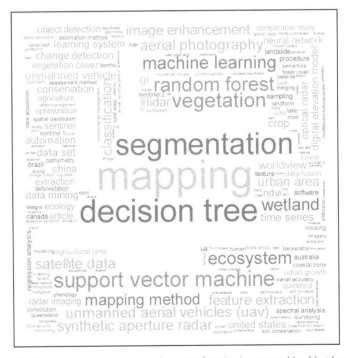

Figure 1. Wordcloud showing the frequently covered topics in geographic object-based image analysis (GEOBIA).

As another way of looking at the recent areas of interest in GEOBIA research, we also identified the papers that were most frequently cited in these 369 articles (Table 1). Aside from review articles covering the field as a whole [4,11], the remainder of the 10 most frequently cited papers all dealt with image segmentation parameter selection [12–14] or image classification [15–18]/change detection [7]. This is similar to the result of the title/keyword/abstract text analysis, and indicates that the general areas of interest within GEOBIA are still related to image segmentation and classification/mapping of land-use/land-cover objects of interest.

Table 1. Ten most frequently cited papers in recent articles on GEOBIA (based on an analysis of 369 articles published in Scopus indexed journals from 2018–2019), and the focus of each paper.

Paper Title	# of Times Cited	Year of Publication	Focus of Paper
Object based image analysis for remote sensing [4]	74	2010	Review
Per-pixel vs. object-based classification of urban land cover extraction using high spatial resolution imagery [15]	39	2011	Image classification
Unsupervised image segmentation evaluation and refinement using a multi-scale approach [12]	34	2011	Segmentation parameter selection
Geographic object-based image analysis–towards a new paradigm [11]	34	2014	Review
A review of supervised object-based land-cover image classification [16]	30	2017	Image classification
Change detection from remotely sensed images: From pixel-based to object-based approaches [7]	23	2013	Change detection
An assessment of the effectiveness of a random forest classifier for land-cover classification [18]	22	2012	Image classification
Automated parameterisation for multi-scale image segmentation on multiple layers [13]	22	2014	Segmentation parameter selection
Discrepancy measures for selecting optimal combination of parameter values in object-based image analysis [14]	20	2012	Segmentation parameter selection
Training set size, scale, and features in Geographic Object-Based Image Analysis of very high resolution unmanned aerial vehicle imagery [17]	20	2015	Image classification

2.2. Special Issue of Remote Sensing on "Image Segmen Tation for Environmental Monitoring"

In December 2019, a Special Issue on the topic of GEOBIA was published in *Remote Sensing*, entitled "Image segmentation for environmental monitoring". The eight papers published in the special issue were largely representative of the current topics of interest within GEOBIA, covering image segmentation algorithm development [19,20] and segmentation parameter optimization strategies [21,22] as well as object-based image classification [23–25] and image fusion [26] methods.

On the topic of image segmentation algorithm development, Tang et al. [19] proposed a nonparametric clustering-based segmentation approach called the edge dependent Chinese restaurant process (EDCRP) method, which utilizes both spectral and spatial information for segmentation, and has the benefit of automatically determining the appropriate number of segments to generate. The EDCRP method was found to produce more accurate segmentation results than several other state-of-the-art methods, although it was more computationally intensive. On the other hand, Shepherd et al. [20] proposed a fast clustering-based approach which uses k-means clustering to generate initial clusters of pixels, followed by a local elimination procedure to aggregate small clusters of pixels until a predefined minimum mapping unit size is met. The high speed and scalability of this approach allowed it to be used to segment a mosaic image of the entire continent of Australia at 30m resolution. Notably, a downloadable tool for implementing this method was made available by the authors at https://www.rsgislib.org/.

On the topic of image segmentation parameter selection/optimization, Georganos et al. [21] and Xiao et al. [22] both developed new methods for local (as opposed to global) optimization of segmentation parameters. Georganos et al. [21] approached the problem by first sub-dividing a study area image into smaller sub-regions, and then performing parameter optimization for each of these sub-regions separately. On the other hand, Xiao et al. [22] first identified globally-optimal segmentation parameters, and then refined this initial segmentation to better delineate different types of urban greenery, by utilizing local information (mean pixel values and standard deviation values within each initial segment). Both of these local approaches were found to outperform global segmentation parameter optimization approaches.

On the topic of object-based image classification, Roodposhti et al. [24] developed a robust rule-based ensemble framework (dictionary of trusted rules, or DoTRules) based on mean-shift segmentation. The approach was tested on three common hyperspectral image benchmark datasets, and found to outperform other ensemble classifiers and support vector machines in many cases. Samat et al. [23] mapped vegetation types in an arid landscape, utilizing an object-based morphological profile method ("extended object-guided morphological profile") to extract contextual features and ensemble algorithms for classification. Finally, Lu et al. [25] applied popular deep learning and transfer learning methods in an object-based image analysis framework to detect landslides in UAV images.

On the last topic, image fusion, Radoux et al. [26] focused on the topic of ecotope mapping using a GEOBIA workflow and multisensor remote sensing data. They found that fusion of aerial optical imagery (blue, green, red, and near-infrared bands) and Lidar topographic data (digital height model and hillshade maps) improved the automated delineation of ecotopes (the smallest ecologically distinct features in a landscape classification system [26]).

We were delighted to receive many high quality papers for this special issue, and would like to sincerely thank all of the authors who submitted their work.

3. Researchers' Views on the Current Status and Future Priorities of GEOBIA

As a final effort of this Special Issue, we disseminated an online questionnaire to the corresponding authors of journal articles published on the topic in the last two years (i.e., the corresponding authors of the 369 journal articles we found in Scopus), and compiled all of the authors' responses (Table S1). Table 2 shows the questions asked in the survey.

Invitations to participate in the survey were sent by email in March 2020, and we received 45 responses in total. The number of years that the respondents had been using GEOBIA approaches (Q1) ranged from 1–20, with an average of 7.18 years (Figure 2). Around half (46%) of the respondents reported that they used GEOBIA approaches more frequently than other remote sensing image analysis approaches, and another 40% used them about as frequently as other approaches (Q2) (Figure 3). The responses to these two questions suggest that survey respondents were generally quite experienced in the use of GEOBIA.

Among the topics within GEOBIA that were currently not receiving sufficient research attention (Q3), object-based accuracy assessment was the most frequently noted (by 22 respondents), followed by big image data analysis (indicated by 19 respondents), and multi-sensor/multi-temporal data fusion (indicated by 17 respondents) (Figure 4). The latter two topics may be particularly important in the context of the growing archives of free high and moderate spatial resolution satellite data provided by different countries' space programs. Among the types of environments that were currently not receiving sufficient research attention (Q4), post-disaster areas was the most frequently indicated (by 18 respondents), followed by coastal areas (indicated by 14 respondents) (Figure 5). Interestingly, urban/built-up areas were least frequently indicated for this question, suggesting a potential oversaturation of urban GEOBIA studies. Finally, in response to Q5, the majority of respondents perceived the current image segmentation and GEOBIA approaches as already having received a relatively high level of maturity (i.e., value of 7 or 8 on a scale from 1 ("They are still at a very early stage of development") to 10 ("They are already good enough, and little-to-no further improvements are required.")) (Figure 6). That said, several remaining weaknesses of GEOBIA were pointed out in response to Q6.

Table 2. Questions asked in online survey on image segmentation and GEOBIA.

Question	Format of Response
Q1: How many years have you been using image segmentation and GEOBIA approaches for remote sensing image analysis?	Numerical (1–20)
Q2: How often do you currently use image segmentation/GEOBIA approaches for remote sensing image analysis, compared to other approaches?	Multiple choice
Q3: What topic(s) are, in your opinion, currently NOT receiving sufficient research attention within the field of image segmentation and GEOBIA? (Check all that apply)	Selected from a list (selecting "Other" allows a free response)
Q4: What types of environments are, in your opinion, currently NOT receiving enough research attention within the field of image segmentation and GEOBIA? (Check all that apply)	Selected from a list (selecting "Other" allows a free response)
Q5: On a scale from 1–10, how mature do you believe the current image segmentation and GEOBIA approaches are for remote sensing image analysis?	Numerical score between 1 ("They are still at a very early stage of development") and 10 ("They are already good enough, and little-to-no further improvements are required").
Q6: What do you feel is the biggest remaining weakness of the current image segmentation/GEOBIA approaches? (Up to ~100 words)	Free response
Q7: What, in your opinion, should be a priority for image segmentation and GEOBIA research over the next 5–10 years for the field to further mature? (Up to ~100 words)	Free response

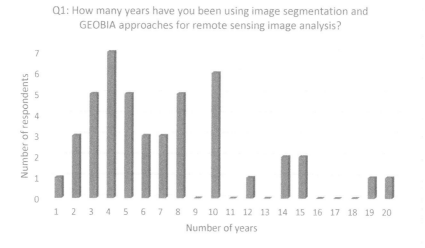

Figure 2. Responses to question 1 (Q1) of the online survey.

Q2: How often do you currently use image segmentation/GEOBIA approaches for remote sensing image analysis, compared to other approaches?

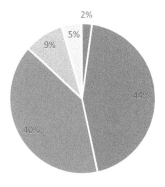

- I use them exclusively, and do not use other types of image analysis approaches. (n = 1)

- I use them more frequently than other types of image analysis approaches. (n = 20)

- I use them about as often as other types of image analysis approaches

- I use them less frequently than other types of image analysis approaches.

Figure 3. Responses to question 2 (Q2) of the online survey.

Q3: What topic(s) are, in your opinion, currently NOT receiving sufficient research attention within the field of image segmentation and GEOBIA? (Check all that apply)

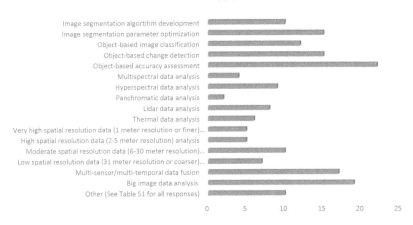

Figure 4. Responses to question 3 (Q3) of the online survey.

The replies to the free response questions on the biggest remaining weaknesses (Q6) and future priorities (Q7) of image segmentation and GEOBIA research are all included in Table S1, and intended to serve as the respondents' anonymous messages to the GEOBIA community. Various views were expressed in response to these two questions, but some common responses were that there is a need for:

- More free (and user-friendly) GEOBIA software and tools;
- Further automation of the segmentation process (especially the parameter setting process);
- More efficient algorithms for handling large image datasets (e.g., for regional/global scale analyses, hyperspectral image analysis, or time-series image analysis);
- Better integration of GEOBIA with deep learning methods as well as 3-D image data;
- More suitable/more standardized accuracy assessment methods.

Some of the other views expressed were unique and quite thought provoking. One interesting response to Q7 stressed the need for greater inclusiveness and creativity, as "Right now the domain as a whole is very centrally controlled by a few people who have clout, and there should be more room for creative ideas." Another interesting response to Q7 was that GEOBIA research should put more attention on "Detecting individual animals from high spatial resolution imagery". Most GEOBIA research to date has focused on detection of land features or artificial features of interest, but expanding its applicability to animal monitoring could help broaden interest in GEOBIA. Although there is not space to highlight all of the other responses to the survey (see Table S1), we hope they can provide some general ideas for future GEOBIA research.

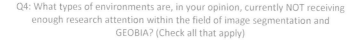
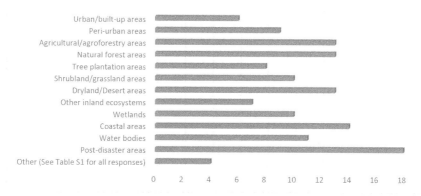

Figure 5. Responses to question 4 (Q4) of the online survey.

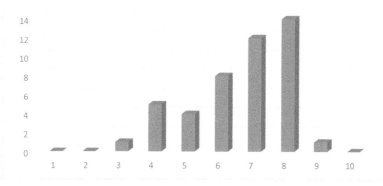

Figure 6. Responses to question 5 (Q5) of the survey. Values range from 1-10, with a value of 1 indicating a respondent perceived that "They are still at a very early stage of development", and a value of 10 indicating the respondent perceived that "They are already good enough, and little-to-no further improvements are required".

To conclude this Special Issue Editorial, we would like to again express our sincere thanks to all of the authors who submitted their work, and to all of the researchers who responded to our questionnaire survey. Much has been accomplished in the first two decades of GEOBIA research, and we look forward to the new developments the next two will bring!

Supplementary Materials: The following are available online at http://www.mdpi.com/2072-4292/12/11/1772/s1, Table S1: Responses to online questionnaire survey on GEOBIA.

Author Contributions: Conceptualization, B.A.J. and L.M.; methodology, B.A.J. and L.M.; software, L.M.; formal analysis, B.A.J. and L.M.; investigation, B.A.J. and L.M.; writing—original draft preparation, B.A.J. and L.M.; writing—review and editing, B.A.J. and L.M.; visualization, B.A.J. and L.M.; supervision, B.A.J. All authors have read and agreed to the published version of the manuscript.

Funding: This research was supported by the Environment Research and Technology Development Fund (S-15-1(4)) Predicting and Assessing Natural Capital and Ecosystem Services (PANCES)) of the Japanese Ministry of the Environment.

Acknowledgments: Thank you to all of the authors who submitted their work to this special issue, and to all of the researchers who responded to our questionnaire survey.

Conflicts of Interest: The authors declare no conflict of interest. The funders had no role in the design of the study; in the collection, analyses, or interpretation of data; in the writing of the manuscript, or in the decision to publish the results.

References

1. Hay, G.J.; Castilla, G. Geographic object-based image analysis (GEOBIA): A new name for a new discipline. In *Object-Based Image Analysis*; Blaschke, T., Lang, S., Hay, G., Eds.; Springer: Berlin/Heidelberg, Germany; New York, NY, USA, 2008; pp. 75–89.
2. Baatz, M.; Schape, A. Multiresolution segmentation—An optimization approach for high quality multi-scale image segmentation. In *Angewandte Geographische Informations-Verarbeitung XII.*; Strobl, J., Blaschke, T., Griesebner, G., Eds.; Wichmann Verlag: Karlsruhe, Germany, 2000; pp. 12–23.
3. Blaschke, T.; Lang, S.; Lorup, E.; Strobl, J.; Zeil, P. Object-oriented image processing in an integrated GIS/remote sensing environment and perspectives for environmental applications. Environmental information for planning, politics and the public. In *Environmental information for Planning, Politics and the Public*; Cremers, A., Greve, K., Eds.; Metropolis Verlag: Marburg, Germany, 2000; Volume 2, pp. 555–570. ISBN 3895183075.
4. Blaschke, T. Object based image analysis for remote sensing. *ISPRS J. Photogramm. Remote Sens.* **2010**, *65*, 2–16. [CrossRef]
5. Chen, G.; Weng, Q.; Hay, G.J.; He, Y. Geographic object-based image analysis (GEOBIA): Emerging trends and future opportunities. *GIScience Remote Sens.* **2018**, *55*, 159–182. [CrossRef]
6. Hossain, M.D.; Chen, D. Segmentation for Object-Based Image Analysis (OBIA): A review of algorithms and challenges from remote sensing perspective. *ISPRS J. Photogramm. Remote Sens.* **2019**, *150*, 115–134. [CrossRef]
7. Hussain, M.; Chen, D.; Cheng, A.; Wei, H.; Stanley, D. Change detection from remotely sensed images: From pixel-based to object-based approaches. *ISPRS J. Photogramm. Remote Sens.* **2013**, *80*, 91–106. [CrossRef]
8. Chen, C. CiteSpace II: Detecting and Visualizing Emerging Trends and Transient Patterns in Scientific Literature. *J. Am. Soc. Inf. Sci. Technol.* **2006**, *57*, 359–377. [CrossRef]
9. Breiman, L. Random Forests. *Mach. Learn.* **2001**, *45*, 5–32. [CrossRef]
10. Cortes, C.; Vapnik, V. Support-Vector Networks. *Mach. Learn.* **1995**, *20*, 273–297. [CrossRef]
11. Blaschke, T.; Hay, G.J.; Kelly, M.; Lang, S.; Hofmann, P.; Addink, E.; Queiroz Feitosa, R.; van der Meer, F.; van der Werff, H.; van Coillie, F.; et al. Geographic Object-Based Image Analysis - Towards a new paradigm. *ISPRS J. Photogramm. Remote Sens.* **2014**, *87*, 180–191. [CrossRef]
12. Johnson, B.A.; Xie, Z. Unsupervised image segmentation evaluation and refinement using a multi-scale approach. *ISPRS J. Photogramm. Remote Sens.* **2011**, *66*, 473–483. [CrossRef]
13. Drăguţ, L.; Csillik, O.; Eisank, C.; Tiede, D. Automated parameterisation for multi-scale image segmentation on multiple layers. *ISPRS J. Photogramm. Remote Sens.* **2014**, *88*, 119–127. [CrossRef]

14. Liu, Y.; Bian, L.; Meng, Y.; Wang, H.; Zhang, S.; Yang, Y.; Shao, X.; Wang, B. Discrepancy measures for selecting optimal combination of parameter values in object-based image analysis. *ISPRS J. Photogramm. Remote Sens.* **2012**, *68*, 144–156. [CrossRef]

15. Myint, S.W.; Gober, P.; Brazel, A.; Grossman-Clarke, S.; Weng, Q. Per-pixel vs. object-based classification of urban land cover extraction using high spatial resolution imagery. *Remote Sens. Environ.* **2011**, *115*, 1145–1161. [CrossRef]

16. Ma, L.; Li, M.; Ma, X.; Cheng, L.; Du, P.; Liu, Y. A review of supervised object-based land-cover image classification. *ISPRS J. Photogramm. Remote Sens.* **2017**, *130*, 277–293. [CrossRef]

17. Ma, L.; Cheng, L.; Li, M.; Liu, Y.; Ma, X. Training set size, scale, and features in Geographic Object-Based Image Analysis of very high resolution unmanned aerial vehicle imagery. *ISPRS J. Photogramm. Remote Sens.* **2015**, *102*, 14–27. [CrossRef]

18. Rodriguez-Galiano, V.F.; Ghimire, B.; Rogan, J.; Chica-Olmo, M.; Rigol-Sanchez, J.P. An assessment of the effectiveness of a random forest classifier for land-cover classification. *ISPRS J. Photogramm. Remote Sens.* **2012**, *67*, 93–104. [CrossRef]

19. Tang, H.; Zhai, X.; Huang, W. Edge Dependent Chinese restaurant process for Very High Resolution (VHR) satellite image over-segmentation. *Remote Sens.* **2018**, *10*, 1519. [CrossRef]

20. Shepherd, J.; Bunting, P.; Dymond, J. Operational Large-Scale Segmentation of Imagery Based on Iterative Elimination. *Remote Sens.* **2019**, *11*, 658. [CrossRef]

21. Georganos, S.; Grippa, T.; Lennert, M.; Vanhuysse, S.; Johnson, B.A.; Wolff, E. Scale matters: Spatially Partitioned Unsupervised Segmentation Parameter Optimization for large and heterogeneous satellite images. *Remote Sens.* **2018**, *10*, 1440. [CrossRef]

22. Xiao, P.; Zhang, X.; Zhang, H.; Hu, R.; Feng, X. Multiscale optimized segmentation of urban green cover in high resolution remote sensing image. *Remote Sens.* **2018**, *10*, 1813. [CrossRef]

23. Samat, A.; Yokoya, N.; Du, P.; Liu, S.; Ma, L.; Ge, Y.; Issanova, G.; Saparov, A.; Abuduwaili, J.; Lin, C. Direct, ECOC, ND and END frameworks-which one is the best? An empirical study of Sentinel-2A MSIL1C image classification for arid-land vegetation mapping in the Ili River delta, Kazakhstan. *Remote Sens.* **2019**, *11*, 1953. [CrossRef]

24. Roodposhti, M.S.; Lucieer, A.; Anees, A.; Bryan, B.A. A robust rule-based ensemble framework using mean-shift segmentation for hyperspectral image classification. *Remote Sens.* **2019**, *11*, 2057. [CrossRef]

25. Lu, H.; Ma, L.; Fu, X.; Liu, C.; Wang, Z.; Tang, M.; Li, N. Landslides information extraction using object-oriented image analysis paradigm based on deep learning and transfer learning. *Remote Sens.* **2020**, *12*, 752. [CrossRef]

26. Radoux, J.; Bourdouxhe, A.; Coos, W.; Dufrêne, M.; Defourny, P. Improving ecotope segmentation by combining topographic and spectral data. *Remote Sens.* **2019**, *11*, 354. [CrossRef]

remote sensing

Article

Scale Matters: Spatially Partitioned Unsupervised Segmentation Parameter Optimization for Large and Heterogeneous Satellite Images

Stefanos Georganos [1,*], Tais Grippa [1], Moritz Lennert [1], Sabine Vanhuysse [1],
Brian Alan Johnson [2] and Eléonore Wolff [1]

[1] Department of Geosciences, Environment & Society, Université libre de Bruxelles (ULB), 1050 Bruxelles, Belgium; tgrippa@ulb.ac.be (T.G.); mlennert@ulb.ac.be (M.L.); svhuysse@ulb.ac.be (S.V.); ewolff@ulb.ac.be (E.W.)
[2] Natural Resources and Ecosystem Services Area, Institute for Global Environmental Strategies, 2108-11 Kamiyamaguchi, Hayama, Kanagawa 240-0115, Japan; johnson@iges.or.jp
[*] Correspondence: sgeorgan@ulb.ac.be; Tel.: +32-2-650-6806

Received: 13 August 2018; Accepted: 5 September 2018; Published: 9 September 2018

check for
updates

Abstract: To classify Very-High-Resolution (VHR) imagery, Geographic Object Based Image Analysis (GEOBIA) is the most popular method used to produce high quality Land-Use/Land-Cover maps. A crucial step in GEOBIA is the appropriate parametrization of the segmentation algorithm prior to the classification. However, little effort has been made to automatically optimize GEOBIA algorithms in an unsupervised and spatially meaningful manner. So far, most Unsupervised Segmentation Parameter Optimization (USPO) techniques, assume spatial stationarity for the whole study area extent. This can be questionable, particularly for applications in geographically large and heterogeneous urban areas. In this study, we employed a novel framework named Spatially Partitioned Unsupervised Segmentation Parameter Optimization (SPUSPO), which optimizes segmentation parameters locally rather than globally, for the Sub-Saharan African city of Ouagadougou, Burkina Faso, using WorldView-3 imagery (607 km^2). The results showed that there exists significant spatial variation in the optimal segmentation parameters suggested by USPO across the whole scene, which follows landscape patterns—mainly of the various built-up and vegetation types. The most appropriate automatic spatial partitioning method from the investigated techniques, was an edge-detection cutline algorithm, which achieved higher classification accuracy than a global optimization, better predicted built-up regions, and did not suffer from edge effects. The overall classification accuracy using SPUSPO was 90.5%, whilst the accuracy from undertaking a traditional USPO approach was 89.5%. The differences between them were statistically significant ($p < 0.05$) based on a McNemar's test of similarity. Our methods were validated further by employing a segmentation goodness metric, Area Fit Index (AFI)on building objects across Ouagadougou, which suggested that a global USPO was more over-segmented than our local approach. The mean AFI values for SPUSPO and USPO were 0.28 and 0.36, respectively. Finally, the processing was carried out using the open-source software GRASS GIS, due to its efficiency in raster-based applications.

Keywords: unsupervised segmentation parameter optimization; GRASS GIS; image classification; land cover; urban areas; big data

1. Introduction

Accurate and precise Land-Use/Land-Cover (LULC) maps derived from remotely sensed imagery are crucial for applications spanning several fields, including spatial planning, population estimation, environmental monitoring, and socio-economic and epidemiological modelling [1–4]. These map products not only provide useful information on their own, but also through their use as an input to secondary models (e.g., population distribution models [3], hydrological models [5], or LULC change models [6–8]. As such, maximizing the accuracy of LULC maps is a critical methodological facet in reducing error propagation and enhancing the effectiveness of science-based policy-making.

For the classification of high- and very-high resolution (VHR) imagery in particular, Geographic Object-Based Image (GEOBIA) analysis has been established as a superior method over traditional pixel-based approaches [9], as it produces a semantic representation of data closer to reality than the arbitrary nature of pixels [10]. Recent studies have attempted to establish a formal ontological framework to further advance the use of objects as spatial representation units [11]. In GEOBIA, the most crucial step before classification is the clustering of neighboring image pixels into segments based on spatial, spectral, and contextual criteria [12]. These segments should ideally represent real world objects or LC categories (e.g., building rooftops, or agricultural fields) that are larger than the original image resolution [13]. As several studies have demonstrated, GEOBIA classification accuracy is not only affected by the classification algorithm itself [14], but also by the quality of the extracted image segmentation [15–18]. Consequently, the selection of an appropriate segmentation (i.e., object-creating) algorithm, as well as its parametrization, are crucial with respect to the final output [19–21].

Region-growing (RG) segmentation techniques are the most popular in GEOBIA literature, mainly due to their implementation through the multiresolution segmentation algorithm [22], implemented in the popular software eCognition (Definiens) [16,23–26]. The most important parameter in RG segmentation is the Threshold Parameter (TP; e.g., the Scale Parameter of the multiresolution segmentation algorithm in eCognition), which governs the average size of the created segments. The selection of the parameter is most commonly attempted through a time consuming, user dependent, trial and error process [27,28], in which the quality of the produced segmentations is assessed visually [29], or through a quantitative comparison against reference data (i.e., manually digitized polygons based on visual image interpretation) [30–32]. These approaches have been criticized for being untenable due to their subjective nature and time inefficiency, whilst at the same time, the improvement they can offer in classification accuracy might be limited [33]. Therefore, other research has been directed towards the development of objectively defined Unsupervised Segmentation Parameter Optimization (USPO) techniques, which evaluate individual segmentations based on geostatistical metrics and do not require reference data [34–36]. To do so, various USPO metrics have been proposed, such as the rate of change in local variance implemented through the estimation of scale parameter tool (ESP) [34,37], the optimization of objective functions such as the Global Score (GS) [38] and the F-measure [18,39] among others, with varying degrees of success. In the comparative study of Grybas et al. [23], the F-measure was found superior to the ESP and GS, potentially due to its sensitivity to over and under segmentation. The GS and F-measure assess spectral values within (i.e., Weighted Variance (WV)) and between (i.e., Global Moran's I (MI)) segments. Ideally, an accurate segmentation should minimize the spectral heterogeneity within segments and maximize the spectral heterogeneity between segments, so the TP that is found to maximize the aforementioned function is accredited to be optimal [40].

So far, the optimization of segment-creating algorithms (and in this study, the region growing one), has been attempted mainly through the use of global methods, either at single or multiple scales [36,37]. A global approach implies that the optimization of the TP is adequate using the whole extent of the study area or a subset which is assumed to be representative [15,33,41]. The vast majority of the developments in the past years operate on that assumption, a situation exaggerated from the relatively small study areas that are used (<3 km^2 in ~95% of the recent literature on object-based land cover mapping) as pointed out in the review of Ma et al. [42]. These approaches assume spatial stationarity—that the relationship between input data and the segment generating process is stable across space which is reflected by having a spatially invariant TP for the whole study area. However, this begs the question "Why is the extent of the study area in a remote sensing application automatically assumed to be the most appropriate scale to optimize the segmentation algorithms?". This is of increasing importance as it has been recently demonstrated that partitioning the study area in smaller regions can provide significantly different results, highlighting the effect of geographic scale in remote sensing operations [43,44]. Spatial stationarity might hold for small homogenous regions, but perhaps is unsuitable for large and/or heterogenous scenes. It would be sensible to hypothesize that the optimal TP would intrinsically and significantly vary across space due to local variations in data structure, particularly for urban areas, which are known for their landscape variability even within the same LULC class. If a global approach would be used in such a case, it might only capture an average and potentially misleading impression of the situation and lead towards adding bias to the segmentation model, which could be reflected both in segmentation evaluation metrics and classification accuracy. In recent years, few studies have tackled this issue by employing more localized or regionalized procedures.

Johnson and Xie [36] refined their global segmentation results in a two-stage procedure by re-segmenting local outliers using geospatial metrics such as Local Moran's I. Cánovas-García and Alonso-Sarría [43] demonstrated improvements in segmentation quality by optimizing the TP independently in agricultural plots, instead of selecting a single parameter for the whole dataset. However, the spatial units were selected a priori by using land use parcel vectors, which requires ancillary data and expert knowledge of the study area. Recently, Kavzoglu et al. [35] proposed a regionalized multiscale approach for small, semi-urban environments where initial, broad scale segments derived from the coarse segmentation selection of the ESP tool, defined further areas for calibrating segmentation parameters. Classification results were shown to improve as the parametrization of the TP was performed regionally, rather than globally. The improvement local methods offer for urban LULC mapping has been recently demonstrated by Grippa et al. [44], where the study area was manually delineated into morphological zones that share similar built-up characteristics, and a separate USPO optimization was applied to each one of them. Nonetheless, the operational capabilities of such methods are either restricted computationally or require tedious manual labor and user expertise that is rarely available. These limitations are important given the advent of big data, which includes the use of VHR datasets at an increasing pace [45]. As such, our effort focuses on semi-automatically identifying and quantifying the degree of spatial non-stationarity and geographic scale dependency between the algorithm parameters for large and heterogeneous satellite images [1].

Our main hypothesis questions the use of global methods a priori, when heterogeneous and/or large datasets are employed. To do so, a discrimination between the observation and operating scales between the TP and USPO optimization must be made. The observation scale corresponds to the whole extent of the study area, whilst the operating scale can be a spatial delineation, which better reflects the optimization of a segmentation algorithm. In simpler words, we are asking the question: "Are the segmentation results better if we analyze the data locally rather than globally?".

In this paper, we present a methodological framework named Spatially Partitioned Unsupervised Segmentation Parameter Optimization (SPUSPO) in which optimization of the TP is performed in a localized manner. The proposed methods are automated and do not require reference information. The underlying rationale of SPUSPO is based on the first law of geography [46] that "all things are related but near things are more related", which suggests that objects being near each other (e.g., built-up characteristics of a neighborhood) have a higher degree of similarity than a set of objects far away. The results of the local optimizations are analyzed, mapped and quantified through spatial statistics, highlighting the variation of segmentation parameters as a function of location and spatial scale. The presented methods are evaluated both at the segmentation and classification level. As a proof of concept, we evaluated the procedure for the large, heterogenous city of Ouagadougou, capital of Burkina Faso. All of the analysis was performed using the GRASS open source GIS software along with open access processing chains suited for satellite VHR datasets [47].

2. Materials and Methods

2.1. Study Area and Data

The study area covered the city of Ouagadougou, capital city of Burkina Faso in Sub-Saharan Africa (SSA). Ouagadougou comprises a complex and heterogenous urban landscape of planned and unplanned neighborhoods and buildings, of various sizes and materials [48]. The city has been undergoing extensive and partly unregulated urban growth (i.e., rural to urban migration) over the last decades [49,50]. To map the LULC of the city, we used a 4-band (R, G, B, NIR) WorldView-3 multispectral image (607 km², Figure 1) from October 2015, and a normalized Digital Surface Model (nDSM) derived from stereo image acquisitions on the same image date. The native spatial resolution of the Worldview-3 imagery is 0.30 cm but was resampled at 0.50 cm by the provider. The value of the elevation information was critical, as the built-up characteristics were very hard to visually discriminate from bare soil and artificial ground surfaces, due to the presence of dust on rooftops and the use of similar construction materials for roofs and artificial ground surfaces. Thus, this challenging study site provided a good stress test for our methods.

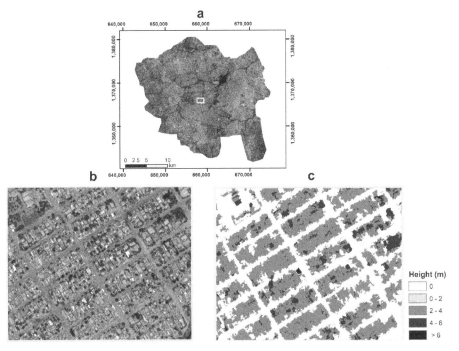

Figure 1. (**a**) Study area extent illustrated from a WorldView-3 RGB composite of Ouagadougou, (**b**) a typical built-up neighborhood of Ouagadougou and (**c**) Normalized Digital Surface Model for the region.

2.2. Segmentation and Unsupervised Segmentation Parameter Optimization

The whole LULC classification framework was realized by employing and extending the semi-automated processing chain proposed by Grippa et al. [1]. The chain was implemented in a Jupyter Notebook format and integrated GRASS GIS functions with Python and R programming languages, framing a complete procedure from the input of initial datasets to final LULC map production. For segmentation, we utilized the RG algorithm implementation of GRASS GIS [51] with all four bands (VNIR) used as inputs. In the GRASS implementation, the TP ranges between 0 to

1, with 0 leading to the situation where each pixel represents a segment, while 1 unifies all image pixels in one object. As Böck et al. [52] pointed out, the USPO metrics are sensitive to the range of candidate segmentations used as input, so we empirically found a range that corresponded to cases of evident over- and under-segmentations to be used as minimum and maximum possible values, as commonly done in similar studies [18,53]. Thus, we evaluated 27 different segmentations starting with a TP of 4 and finishing at a TP of 31, guided by an incrementing step value of 1, as in previous studies, [54]. For reader convenience, all TP values were multiplied by 1000 in the illustrative and text materials.

To evaluate the quality of each of the different segmentations, we used the inter- and intra-segmentation heterogeneity metrics Moran's I (*MI*) and Weighted Variance (*WV*), respectively. *MI* calculates the degree of spatial autocorrelation present in the values of nearby geographic features, and it was used in our case (and in many other USPO studies) to calculate how spectrally heterogeneous segments are, on average, from their neighbors (i.e., in terms of the mean segment values calculated for each spectral band). For this reason, it can provide a measure of "oversegmentation goodness"; Low *MI* values for a segmentation layer indicate low spatial autocorrelation between segment spectral values, suggesting that most segments belong to a different ground feature (with different spectral reflectance properties) than its neighbors. *WV*, on the other hand, describes the average spectral variability within segments (weighted by each segment's area). *WV* can provide a measure of "undersegmentation goodness"; Low *WV* values indicate little internal variation in the spectral properties of segments, suggesting the segment does not contain a mixture of multiple ground features. *MI* and *WV* are given by:

$$WV = \frac{\sum_i^n a_i * v_i}{\sum_i^n a_i} \tag{1}$$

$$MI = \frac{n \sum_i^n \sum_j^n w_{ij} z_i z_j}{M \sum_i^n z_i^2} \tag{2}$$

where for Equation (1), n is the number of segments, v_i is the variance and a_i the area for each segment, while in Equation (2), n is the number of segments, $z_i = x_i - \bar{x}$, $\bar{x}$ is the mean value of segment x, $M = \sum_{i=1}^n \sum_{j=1}^n w_{ij}$ and w_{ij} is the element of the matrix of spatial proximity M, which indicates the spatial connectivity for segments i and j [52,53].

To perform USPO, the oversegmentation and undersegmentation goodness values calculated for each segmentation layer need to be combined into a single value, e.g., through addition [38] or the F-measure [18]. We used the F-measure to combine *MI* and *WV* values in this study, as it was demonstrated to be less sensitive to excessive over- and undersegmentation than other combination approaches in Zhang et al. [39] and implemented in GRASS module "i.segment.uspo" [55]. To derive an F-measure from these two components, we first need to normalize them to a similar range (0–1) [38]:

$$MI_n = \frac{MI_{max} - MI}{MI_{max} - MI_{min}}, \tag{3}$$

$$WV_n = \frac{WV_{max} - WV}{WV_{max} - WV_{min}}, \tag{4}$$

where WV_n is the normalized *WV* (or *MI*), WV_{max} is the maximum *WV* (or *MI*) value of all candidate segmentations, WV_{min} is the minimum *WV* (or *MI*) value of all candidate segmentations and *WV* is the *WV* (or *MI*) value of the current segmentation. The F-measure is a harmonic weighting of these two features:

$$F_{opt} = \left(1 + a^2\right) \frac{WV_{max} - WV}{a^2 * WV_{max} - WV_{min}}, \tag{5}$$

where F_{opt} is the score of a candidate segmentation to be evaluated, ranging from 0 to 1, with higher values indicating higher quality; and a is the relative weight factor that assigns different significance to one metric over the other [18]. In our case we used a relative weight of 1, indicating equal weighting of the *MI* and *WV* components in calculating F_{opt}. The procedures were fully automated and parallelized due to the flexibility of GRASS GIS for applications including large raster datasets.

2.2.1. Global USPO

The conventional global USPO approach involves using either the whole image extent as input to the USPO procedure, or a representative subset [43]. Since our image was very large (20 GB), we used the latter method, as depicted in Figure 2. The selected subset (10 km²) contained planned, unplanned, and industrial built-up zones, with different kinds of vegetation, as well as bare soil, and thus, was deemed an appropriate candidate. The TP resulting from applying USPO to that region was 12, and we consequently used that value to segment all parts of the WorldView-3 image.

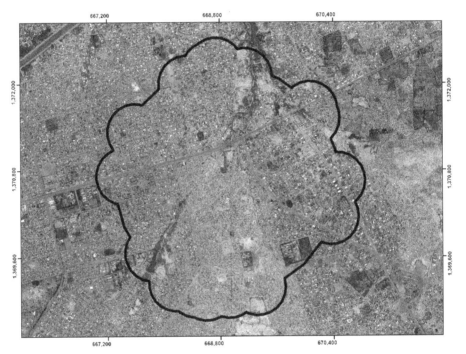

Figure 2. Subset of the WorldView-3 imagery (~10 km²) where the RG's TP was optimized for use in the whole image. The selected area contains a distribution of land cover classes representative of the whole study area.

2.2.2. Spatially Partitioned Unsupervised Segmentation Parameter Optimization (SPUSPO)

As mentioned in the introduction, a global optimization of the USPO might not be appropriate due to the spatial heterogeneity within the image. As such, an alternative approach would be to partition the study area into several subsets, and to apply the optimization procedure locally in each subset. If the segmentation level selected as optimal by a global USPO calculation approach differs significantly from the segmentation level selected locally (i.e., through local USPO calculation in each partition of the study area), a spatially non-stationary process is taking place, and thus a global model might not be the best candidate to use. To investigate this phenomenon, we partitioned the image in three automated ways. The first two methods for partitioning were done using regularly-shaped rectangular tiles of predefined sizes, and the third partitioning method involved automated delineation using a cutline creating algorithm. The predefined partition was based on splitting the WorldView-3 image, into tiles of equal area and for most cases, equal geometry. The area of the rectangular image subsets for the first two partitioning approaches was 0.25 km^2 (P1) and 0.12 km^2 (P2), totaling to 2427 and 4887 subsets, respectively (Figure 3). Although the results of predefined partitioning can be fruitful for exploratory purposes, they suffer from edge effects at their borders. Since they are predefined and fixed in size, they arbitrarily partition the landscape, which can result in noisy/badly segmented objects along the boundaries of the rectangular subsets as artifacts (i.e., splitting building roofs or trees in half). To treat this issue, for the third and main partitioning approach (P3), we deployed a cutline creating algorithm using Laplacian zero-crossing edge detection [56–58], as implemented in the 'i.cutlines' module of GRASS GIS [59]. In that way, the created subsets would delineate the landscape in a more meaningful way, as they would follow linear patterns, such as roof edges and streets. The size of the cutline-created subsets can be decided by the user with respect to the application case. In our case, we created subsets closer to the P2 partition and as such, 4900 subsets were created. Examples of the different spatial partitioning methods are illustrated in Figure 3. In both global and local approaches, the minimum size of a created segment was preset at 14 pixels to avoid unnecessary oversegmentation.

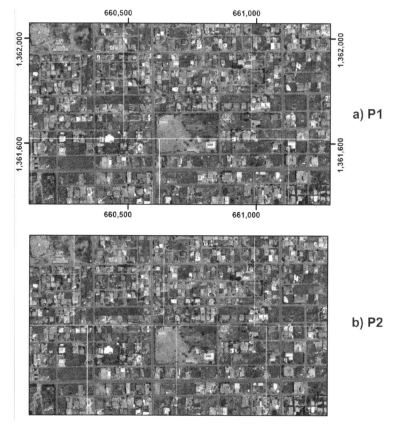

Figure 3. *Cont.*

Figure 3. Partitioning the WorldView-3 image into spatial subsets for local USPO optimization. (a) Delineation by 0.250 km^2 area tiles, (b) delineation by 0.125 km^2 area tiles and (c) delineation based on zero crossing cutline algorithm.

2.2.3. Spatially Partitioned Unsupervised Segmentation Parameter Optimization (SPUSPO)

One of the merits of carrying out a localized approach is that it allows for decomposing a global process, into a wide set of useful information which is mappable. Since USPO was applied locally, a unique TP was produced for each spatial subset. The variation of the local TP from the single TP value of the global USPO can be quantified to assess the degree of spatial non-stationarity. If there would be no unexpected variation in the TP, that would suggest that a global approach is indeed adequate, ceteris paribus. Along with mapping the results, we proposed a Segmentation Parameter Stationarity Index (*SPSI*), which was loosely based on the Stationarity Index of Osborne et al. [60] to assess spatial non-stationarity in gaussian models:

$$SPSI = \frac{IQR(TP_L)}{(TP_G + TP_{step}) - (TP_G - TP_{step})} \qquad (6)$$

where TP_G is the TP of the global USPO, TP_{step} is the step parameter of the USPO procedure, and $IQR(TP_L)$ is the interquartile range of the distribution of the TP's from a local approach. The interquartile range was used to mask outlier TP values that could emerge from random variation. Values equal to or smaller than 1 imply stationarity, as the variation of the local TPs is not exceeding what one would expect from a random process. Values higher than 1 indicate that there is significant spatial variation.

2.3. Land Use and Land Cover Classification

Ultimately, the segments were constructed with the aim of being labeled through a classification model. As such, another method to assess the local and global USPO methods is through the accuracy and performance of a LULC classification. The classification scheme and training data are presented below (Table 1). The training data were collected through random and stratified random sampling, and consisted of 2478 objects across the city, which were labeled through visual interpretation by two experts during the same period. The amount of training data was selected in such way that the addition of new data points did not significantly improve classification accuracy. Swimming pools were sampled manually due to their scarcity. To evaluate the results of the classification between the two methods, we used an expert-based manual delineation of Ouagadougou, based on building size and density [44] (Figure 4). In each one of these built-up types, we randomly sampled 150 points adding up to a total of 1650 points, and computed the Overall Accuracy (OA), as well as the F-score for each LULC class. No overlapping between training and testing data was allowed.

Table 1. Training objects for each LULC class and method.

LULC	Description	Training Set Size
Buildings (BU)		400
Swimming Pool (SP)		179
Artificial Ground Surface (AS)	Asphalt, concrete, semi-built-up constructions	216
Bare Soil (BS)		399
Tree (TR)		191
Low Vegetation (LV)	Grass, bushes, dry vegetation	702
Inland Water (IW)	Lakes, ponds, rivers, wetlands	205
Shadow (SH)		186

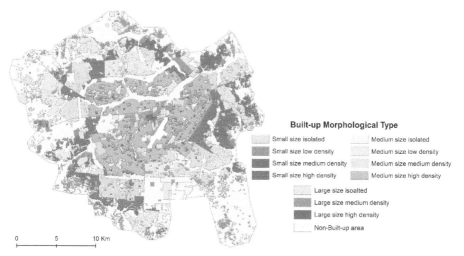

Figure 4. Morphological delineation of Ouagadougou based on built-up size and density categories.

To classify the whole image, we computed several descriptive statistics for segments, based on the values of the pixels located within the segment, i.e., the values of each spectral band, NDVI values, and nDSM values (min, median, mean, max, range, 1st and 3rd quantiles and sum) as well as geometrical covariates (fractal dimension, perimeter, area, compactness). An Extreme Gradient Boosting (XGBoost, R 3.5.1) classifier was used as it was recently shown to outperform benchmark classifiers such as Support Vector Machine in VHR LULC classifications [14]. XGBoost is an ensemble of Classification and Regression Trees that is based in the principle of boosting [61]. The parameters of the algorithm were tuned through Bayesian Optimization [14,62], to ensure the quality of the results. Finally, we performed feature selection to reduce the computational burden and potentially increase the predictive capabilities of the model by deploying the popular Variable Selection with Random Forests (VSURF) algorithm, which is suited for tree-based classifiers such as XGBoost [63,64]. Out of the initial 59 features, 18 were selected by VSURF to build the most discriminant, redundancy-free model.

2.4. Segmentation Goodness Metrics

To evaluate the effect of SPUSPO on the segmentation of buildings, we compared the cutline-based segmentation and the global approach against reference data. In detail, we manually delineated 100 buildings that were randomly selected from the pool of training data used for the LULC classification. Finally, we computed the *Area Fit Index (AFI)* which is a commonly used joint index of over- and undersegmentation [31,32,53]:

$$AFI = \frac{area(x_i) - area(y_{imax})}{area(x_i)} \tag{7}$$

where x_i is the reference object and y_{imax} is the largest relevant segment intersecting x_i. Values closer to 0 suggest a better segmentation, values > 0 imply oversegmentation whereas values < 0 undersegmentation.

2.5. Computational Requirements and Data Availability

The computing infrastructure used for the experiments consisted of two Intel® Xeon® CPU E5-2690 (2 processors of 2.90 GHz, 16 cores, 32 processing threads) and 96 GB of RAM. Segmenting the WorldView-3 image with a single TP parameter (tiled) required roughly 20 h of processing time while on average, a SPUSPO method required about 63 hours by exploiting the parallelization of the 'i.segment.uspo' module of GRASS [56]. The code, results and processed material is openly accessible in the following repository (https://zenodo.org/record/1341116#.W3FSUvZuJ_t) [65].

3. Results

3.1. Threshold Parameter Variation

The spatial variation of the TP was a function of the size and geometry of the subsets used for local optimization. Figure 5 demonstrates that the variation follows patterns of the landscape. The locations where high TP values were selected as optimal were mainly clustered around unplanned, low elevated neighborhoods, whereas the locations where very low TP values were selected as optimal were mostly found in vegetated areas, potentially due to their unique spectral properties (high local variation in the NIR band). The local outputs of each metric used for the local USPO calculations can also be enlightening with respect to illustrating the level of spatial heterogeneity of the imagery. Figures 6 and 7 confirm that MI and WV have an inverse relationship, with MI being decisive in optimization in the central and eastern regions of unplanned areas, and vice-versa. The SPSI value was 1.5 for P1, and 2 for P2 and P3, indicating a non-stationary variation in optimal TP values.

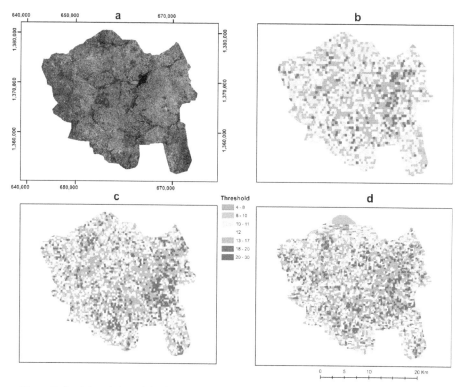

Figure 5. Spatial variation of the threshold parameter (TP) across Ouagadougou. (**a**) WorldView-3 RGB composite, partitioning by (**b**) P1 (**c**) P2 and (**d**) P3 approaches, respectively. The TP controls the average size of the created segments.

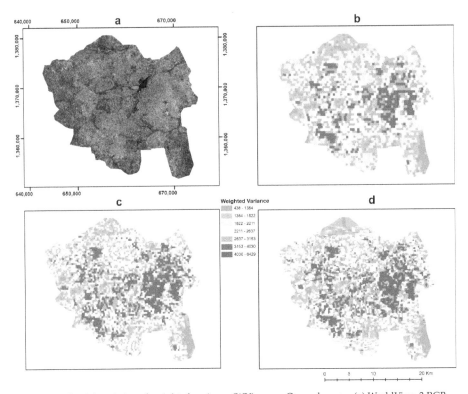

Figure 6. Spatial variation of weighted variance (WV) across Ouagadougou. (**a**) WorldView-3 RGB composite, partitioning by (**b**) P1, (**c**) P2 and (**d**) P3 approaches, respectively. High values of WV indicate large intra-segment variability while low values describe more homogenous objects.

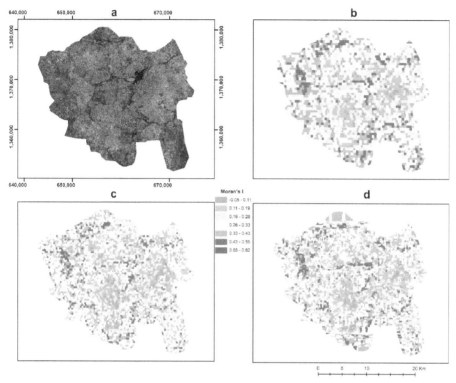

Figure 7. Spatial variation of Moran's I (MI) values across Ouagadougou. (**a**) WorldView-3 RGB composite, partitioning by (**b**) P1 (**c**) P2 and (**d**) P3 approaches, respectively. The higher the MI value, the stronger the effect of spatial autocorrelation between a created segment and its neighbors.

The variability of these parameters was also visualized in a set of boxplots in Figure 8. From this figure, the TP parameter variation is slightly smaller for the P1 approach than for the other two partitioning methods, possible because image partitions of P1 are larger than those of P2 and P3, and thus do not capture as much of the local heterogeneity in urban structure. Notably, when using smaller spatial partitions, MI tends to decrease (and WV tends to increase), which constitute the differences in TP among the different methods.

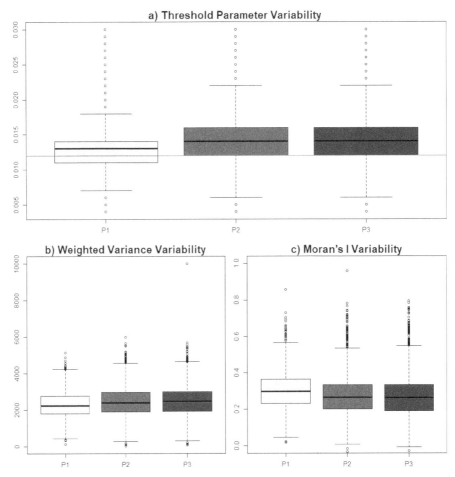

Figure 8. Boxplots demonstrating the variability of (**a**) the TP, (**b**) MI and (**c**) WV for the different partitioning approaches (P1, P2, P3).

3.2. Land-Use Land-Cover Classification

The results of the LULC classification were found to be affected by the segmentation quality. Figures 9 and 10 show case how SPUSPO could enhance classification accuracy by producing segments better fitting the local environment, in various areas in Ouagadougou. Figure 9 demonstrates that in both planned and unplanned regions, the improvement in classification results was mainly due to the cutlines segmentation, delineating the buildings in a less oversegmenting fashion, avoiding overestimation of built-up near the borders due to the inconsistent and "patchy" nature of the nDSM as a predictor, that does not closely follow built-up boundaries.

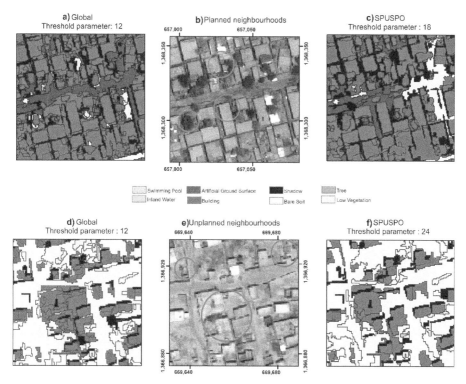

Figure 9. Example of the LULC map classification in a planned and unplanned built-up area. (**a**) LULC classification with a global approach in a planned neighborhood, (**b**) RGB Pleiades Composite, (**c**) LULC classification with a cutline approach in an unplanned neighborhood, and (**d**) LULC classification with a global approach in a planned neighborhood, (**e**) RGB Pleiades Composite, (**f**) LULC classification with a cutline approach in an unplanned neighborhood.

LULC classification based on SPUSPO was superior for vegetation and waterbodies of Ouagadougou. Figure 10 demonstrates cases of confusion between low and high vegetation, when using a global approach. Additionally, the misclassification of water as built-up is significantly less with SPUSPO. Notably, a scene might be segmented with intrinsically different thresholds (Figure 10f), which implies that the reason SPUSPO methods performed better is their incorporation of only the spatial information of the segmented region, and not information that comes from locations far away, which might not be useful at the local level.

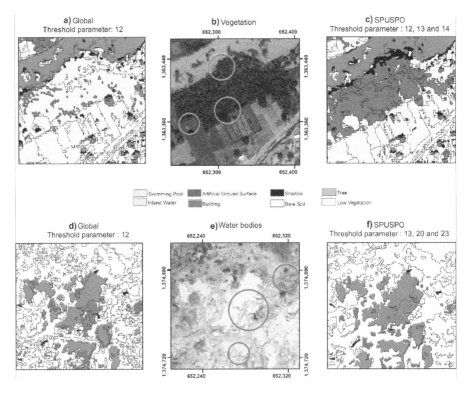

Figure 10. Example of the LULC map classification in a vegetated regions and inland water bodies. (**a**) LULC classification with a global approach in a forested area, (**b**) RGB Pleiades Composite, (**c**) LULC classification with a cutline approach in a forested area, (**d**) LULC classification with a global approach in water bodies, (**e**) RGB Pleiades Composite, and (**f**) LULC classification with a cutline approach in water bodies.

The Overall Accuracy for the SPUSPO and global optimization based on the reference set was 90.5% and 89%, respectively. Moreover, the differences among them were statistically significant, based on a two-tailed McNemar's test of similarity ($p < 0.05$). The local optimization was superior for most cases, both when concerning the OA and per-class evaluation metrics (Table 2). The largest improvements were found in the classification of inland water and shadows (+18% and +3% increase on the F1 score, respectively).

Table 2. Precision, Recall and F-score metrics for each LULC class with SPUSPO and global USPO, respectively.

	Precision		Recall		F1	
Class	SPUSPO	Global	SPUSPO	Global	SPUSPO	Global
Building	0.93	0.93	0.94	0.93	0.94	0.93
Artificial Ground Surface	0.83	0.83	0.88	0.86	0.85	0.84
Bare Soil	0.88	0.84	0.87	0.87	0.88	0.86
Tree	0.81	0.81	0.91	0.93	0.85	0.87
Low veg	0.94	0.94	0.89	0.86	0.91	0.90
Inland Water	0.86	0.73	0.66	0.47	0.75	0.57
Shadow	0.94	0.90	0.95	0.95	0.95	0.92

An additional, indirect way to assess the segmentation quality is to investigate the variable importance of the geometrical covariates. The geometrical covariates that were used in the classification model after VSURF feature selection took place were perimeter, area, and fractal dimension. Figure 11 illustrates the improved effect a local approach has on the importance of most of these variables, further supporting the merit of using SPUSPO. The interpretation of the results, refers to the gain in model accuracy when a feature is used in the splits of the XGBoost tree development. The importance of these covariates is varying, but in all cases, the local approach further enhances their predictive power for classification, since the segments fit better the variability of the local environment.

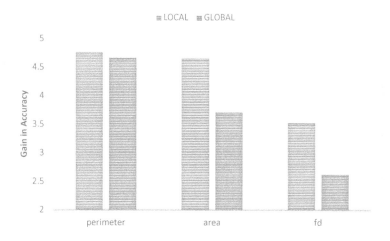

Figure 11. Feature importance of geometrical covariates, as derived from an XGBoost classifier, for the global and cutline segmentation-based approaches, respectively. The method used to derive importance is the gain in accuracy.

3.3. Segmentation Goodness Metrics

The results of the AFI for buildings are depicted in Table 3 through several descriptive statistics. As expected, the building objects were less over segmented with SPUSPO, because the parameter was spatially adapting to characteristics of each built-up neighborhood in Ouagadougou (Figure 12). The AFI values of the local method were consistently closer to zero compared to their counterpart, further promoting the use of this approach.

Table 3. *Area Fit Index* for building objects in Ouagadougou. Values closer to 0 suggest a better segmentation, values > 0 imply over segmentation while values < 0 under segmentation.

Descriptive Statistics	Area Fit Index (*AFI*)	
	SPUSPO	Global
1st	0.04	0.11
Median	0.22	0.38
Mean	0.28	0.36
3rd	0.53	0.62

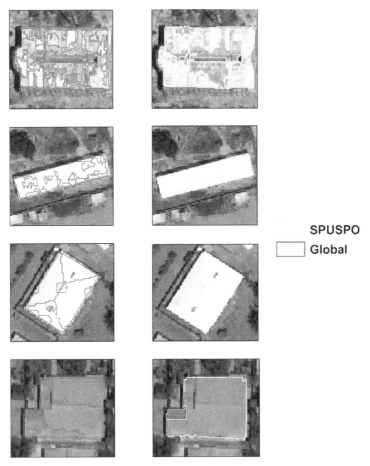

Figure 12. Example segmentations of buildings in Ouagadougou. Red color indicates segments created by a global approach, while green color indicates segments coming from SPUSPO. The decrease of over segmentation is evident in most cases, as the parameters are derived from neighboring locations, better fitting the data structure.

4. Discussion

The results suggested that the benefits of performing SPUSPO, are multiple. To start with, it allows for the local variations in spectral and spatial heterogeneity within an image to be incorporated into the segmentation parameter optimization approach, which is more intuitive because the optimization procedure is derived using the actual locations they are being applied to and not from locations situated afar. This supports the hypothesis that in large and heterogeneous areas, a single TP may be inadequate, as it is simply an average expression of several non-stationary processes. The results confirm prior analysis in another Sub-Saharan city of Dakar, where a semi-automated local approach outperformed classical optimization methods [54]. Moreover, several other studies have described how regionalized approaches can be of merit for urban, semi-rural, and agricultural environments [35,43,44]. Nonetheless, an important facet that has been neglected so far is how to partition the landscape in geographically large areas in conjunction with VHR imagery, and in the absence of reference data such as parcels or blocks. For a continuous LULC map, an appropriate delineation of the image is important, as it must be as adjustable to landscape patterns, such as streets or roofs, as much as possible to avoid/reduce edge effects. Although all local approaches showed they can be of merit, the cutline-based partition helped to specifically address these issues. Undertaking SPUSPO, produced higher classification accuracy than using a traditional global optimization method (+1.5% increase in OA). The results are confirmed further using AFI as a segmentation goodness metric, which showed that building segments from applying SPUSPO are less oversegmented than their global counterparts, with mean values of 0.28 and 0.36 for SPUSPO and global USPO, respectively. The analysis validated our initial hypothesis that the way we look at the data can produce significantly different results, and is related to the importance of appropriate spatial scale selection in geography, which was largely signified through the work of Woodcock and Strahler [66] and Fotheringham et al. [67]. Additionally, a local segmentation optimization approach is not only linked to traditional GEOBIA analysis, but might be needed in large scale applications where deep learning classification is coupled with segments to achieve better object delineation/extraction as demonstrated recently in References [68–70]. Another important piece of information that we can extract from these methods is the ability to map intermediate and final results, which can be enlightening both as a general understanding of how spatial processes operate in the local scale, but also how to calibrate segmentation parameters in further processing if an unsupervised multi-scale framework is selected [18]. The LULC products in SSA cities are often used as inputs for fine scale population modelling, land use, and spatial planning, and consequently, effective policy making, given the extreme scarcity of reference information [2,71,72]. This is significant for the outcome of our analyses because there was better prediction of most classes by the SPUSPO approach; it presents an additional motivation to partake of a local method to reduce error propagation in secondary models.

The main limitation of SPUSPO is the increased computational time and experimentation to detect a satisfactory spatial level to analyze image information, which can vary depending on the image resolution and study area, leading to a trade-off between computational requirements and performance. Therefore, more sophisticated methods are needed to help establish an efficient framework to fully exploit the benefits of local optimization. Ideally, in large and heterogeneous areas, a spatial partition should not suffer from edge effects and should meaningfully delineate the landscape with a certain degree of intra-homogeneity. Cutline partitioning satisfies both criteria to some extent, but its effectiveness can only be determined post-hoc, which increases the computational and time demands as several cutline partitions may need to be evaluated. More adequate methods that can focus in a priori determination of a suitable scale using image statistics, such as spatial dependency among regions [73], could be of benefit to achieve this, particularly in a multi-scale context. Other research should explore the potential of multi-resolution imagery to define operational partitions using top down approaches. For instance, a low-medium resolution LULC product can define homogeneous regions to apply SPUSPO using finer resolution imagery. Moreover, noise additive models could help in better establishing a comparative framework among different segmentation approaches, particularly for SAR or hyperspectral data [74]. A lot of the limitations that come with involving local methods, can be significantly reduced (i) by utilizing GRASS GIS, which is highly parallelized in the USPO optimization module and more notably, performs all the operations in a raster format and does not require vector conversion at any moment, dramatically boosting its effectiveness for large-scale computing; and (ii) invoking state-of-the-art segmentation algorithms, with respect to their computational efficiency, as recently shown by Gu et al. [75].

5. Conclusions

In this study, the optimization of a region-growing segmentation algorithm was attempted using a spatially varying parameter model, named SPUSPO. The whole framework was developed with a focus on automation and large-scale analysis of VHR imagery. The results validated our hypothesis that in large and heterogeneous areas, using only a single set of parameters to optimize the region-growing algorithm was inadequate. Employing as a case study, the city of Ouagadougou, it was demonstrated that undertaking local optimization methods was of merit and led to significantly better LULC classification results (+1.5% increase in OA), validated by a McNemar's test of similarity. Moreover, at the segmentation level, building delineation was improved with a mean Area Fit Index of 0.28 and 0.36 for SPUSPO and global USPO, respectively. Moreover, the feature importance of geometrical covariates is recommended as an indirect measure to assess the quality of a segmentation. We demonstrated that geometrical features were more important and predictive when using local approaches. Finally, GRASS GIS was heavily utilized and is promoted as an open source tool to handle large volumes of data with advanced analysis techniques.

Author Contributions: S.G. wrote the manuscript and performed the analysis. M.L. developed the cutline algorithm. T.G. provided technical assistance and helped develop the experiments. B.J., S.V. and E.W. provided valuable feedback and comments during the internal revisions of the manuscript.

Acknowledgments: This work was supported by BELSPO (Belgian Science Policy Office) in the frame of the STEREO III program—project REACT (SR/00/337). We would also like to thank the three reviewers for their useful and insightful comments and recommendations which significantly improved the quality of the manuscript. WorldView3 data is copyrighted under the mention "©COPYRIGHT 2015 DigitalGlobe, Inc., Longmont CO USA 80503. DigitalGlobe and the DigitalGlobe logos are trademarks of DigitalGlobe, Inc. The use and/or dissemination of this data and/or of any product in any way derived there from are restricted. Unauthorized use and/or dissemination is prohibited".

Conflicts of Interest: The authors declare no conflict of interest.

References

1. Grippa, T.; Lennert, M.; Beaumont, B.; Vanhuysse, S.; Stephenne, N.; Wolff, E. An open-source semi-automated processing chain for urban object-based classification. *Remote Sens.* **2017**, *9*, 358. [CrossRef]
2. Kabaria, C.W.; Molteni, F.; Mandike, R.; Chacky, F.; Noor, A.M.; Snow, R.W.; Linard, C. Mapping intra-urban malaria risk using high resolution satellite imagery: A case study of Dar es Salaam. *Int. J. Health Geogr.* **2016**, *15*, 26. [CrossRef] [PubMed]
3. Linard, C.; Gilbert, M.; Snow, R.W.; Noor, A.M.; Tatem, A.J. Population distribution, settlement patterns and accessibility across Africa in 2010. *PLoS ONE.* **2012**, *7*, e31743. [CrossRef] [PubMed]
4. Taubenbock, H.; Wurm, M.; Setiadi, N.; Gebert, N.; Roth, A.; Strunz, G.; Birkmann, J.; Dech, S. Integrating remote sensing and social science. In Proceedings of the IEEE Joint Urban Remote Sensing Event, Shanghai, China, 20–22 May 2009.
5. Niehoff, D.; Fritsch, U.; Bronstert, A. Land-use impacts on storm-runoff generation: Scenarios of land-use change and simulation of hydrological response in a meso-scale catchment in SW-Germany. *J. Hydrol.* **2002**, *267*, 80–93. [CrossRef]
6. Otukei, J.R.; Blaschke, T. Land cover change assessment using decision trees, support vector machines and maximum likelihood classification algorithms. *Int. J. Appl. Earth Obs. Geoinf.* **2010**, *12*, 27–31. [CrossRef]
7. Manakos, I.; Braun, M. *Land Use and Land Cover Mapping in Europe*, 3rd ed.; Springer Nature: Heidelberg, Germany, 2014.
8. Iizuka, K.; Johnson, B.A.; Onishi, A.; Magcale-Macandog, D.B.; Endo, I.; Bragais, M. Modeling Future Urban Sprawl and Landscape Change in the Laguna de Bay Area, Philippines. *Land* **2017**, *6*, 26. [CrossRef]
9. Blaschke, T.; Hay, G.J.; Kelly, M.; Lang, S.; Hofmann, P.; Addink, E.; Queiroz Feitosa, R.; van der Meer, F.; van der Werff, H.; van Coillie, F.; et al. Geographic Object-Based Image Analysis—Towards a new paradigm. *ISPRS J. Photogramm. Remote Sens.* **2014**, *87*, 180–191. [CrossRef] [PubMed]
10. Chen, G.; Weng, Q.; Hay, G.J.; He, Y. Geographic Object-based Image Analysis (GEOBIA): Emerging trends and future opportunities. *GIScience Remote Sens.* **2018**, *55*. [CrossRef]
11. Gu, H.; Li, H.; Yan, L.; Liu, Z.; Blaschke, T.; Soergel, U. An object-based semantic classification method for high resolution remote sensing imagery using ontology. *Remote Sens.* **2017**, *9*, 329. [CrossRef]
12. Blaschke, T. Object based image analysis for remote sensing. *ISPRS J. Photogramm. Remote Sens.* **2010**, *65*, 2–16. [CrossRef]
13. Rasanen, A.; Rusanen, A.; Kuitunen, M.; Lensu, A. What makes segmentation good? A case study in boreal forest habitat mapping. *Int. J. Remote Sens.* **2013**, *34*, 8603–8627. [CrossRef]

14. Georganos, S.; Grippa, T.; Vanhuysse, S.; Lennert, M.; Shimoni, M.; Wolff, E. Very high resolution object-based land use-land cover urban classification using extreme gradient boosting. *IEEE Geosci. Remote Sens. Lett.* **2018**, *15*, 607–611. [CrossRef]

15. Ma, L.; Li, M.; Blaschke, T.; Ma, X.; Tiede, D.; Cheng, L.; Chen, Z.; Chen, D. Object-based change detection in urban areas: The effects of segmentation strategy, scale, and feature space on unsupervised methods. *Remote Sens.* **2016**, *8*, 761. [CrossRef]

16. Srivastava, M.; Arora, M.K.; Raman, B. Selection of critical segmentation-A prerequisite for Object based image classification. In Proceedings of the 2015 National Conference on Recent Advances in Electronics & Computer Engineering (RAECE), Roorkee, India, 13–15 February 2015.

17. Lowe, S.H.; Guo, X. Detecting an optimal scale parameter in object-oriented classification. *IEEE J. Sel. Top. Appl. Earth Obs. Remote Sens.* **2011**, *4*, 890–895. [CrossRef]

18. Johnson, B.; Bragais, M.; Endo, I.; Magcale-Macandog, D.; Macandog, P. Image segmentation parameter optimization considering within- and between-segment heterogeneity at multiple scale levels: Test case for mapping residential areas using landsat imagery. *ISPRS Int. J. Geo-Inform.* **2015**, *4*, 2292–2305. [CrossRef]

19. Gao, Y.A.N.; Mas, J.F.; Kerle, N.; Navarrete Pacheco, J.A. Optimal region growing segmentation and its effect on classification accuracy. *Int. J. Remote Sens.* **2011**, *32*, 3747–3763. [CrossRef]

20. Yang, J.; Li, P.; He, Y. A multi-band approach to unsupervised scale parameter selection for multi-scale image segmentation. *ISPRS J. Photogramm. Remote Sens.* **2014**, *94*, 13–24. [CrossRef]

21. Zhang, Q.; Huang, X.; Zhang, L. An energy-driven total variation model for segmentation and classification of high spatial resolution remote-sensing imagery. *IEEE Geosci. Remote Sens. Lett.* **2013**, *10*, 125–129. [CrossRef]

22. Baatz, M.; Schape, A. Multiresolution Segmentation: An Optimization Approach for High Quality Multi-Scale Image Segmentation. 2000. Available online: https://www.semanticscholar. org/paper/Multiresolution-Segmentation-an-optimization-appro-Baatz-Sch%C3%A4pe/ 364cc1ff514a2e11d21a101dc072575e5487d17e (accessed on 20 December 2017).

23. Grybas, H.; Melendy, L.; Congalton, R.G. A comparison of unsupervised segmentation parameter optimization approaches using moderate- and high-resolution imagery. *GISci. Remote Sens.* **2017**, *54*, 515–533. [CrossRef]

24. Du, S.; Guo, Z.; Wang, W.; Guo, L.; Nie, J. A comparative study of the segmentation of weighted aggregation and multiresolution segmentation. *GISci. Remote Sens.* **2016**, *53*, 1–20. [CrossRef]

25. Mesner, N.; Oštir, K. Investigating the impact of spatial and spectral resolution of satellite images on segmentation quality. *J. Appl. Remote Sens.* **2014**, *8*, 83696. [CrossRef]

26. Zhong, Y.; Gao, R.; Zhang, L. Multiscale and multifeature normalized cut segmentation for high spatial resolution remote sensing imagery. *IEEE Trans. Geosci. Remote Sens.* **2016**, *54*, 6061–6075. [CrossRef]

27. Duro, D.C.; Franklin, S.E.; Dube, M.G. A comparison of pixel-based and object-based image analysis with selected machine learning algorithms for the classification of agricultural landscapes using SPOT-5 HRG imagery. *Remote Sens. Environ.* **2012**, *118*, 259–272. [CrossRef]

28. Zhang, H.; Fritts, J.E.; Goldman, S.A. Image segmentation evaluation: A survey of unsupervised methods. *Comput. Vis. Image Underst.* **2008**, *110*, 260–280. [CrossRef]

29. Flanders, D.; Hall-Beyer, M.; Pereverzoff, J. Preliminary evaluation of ecognition object-based software for cut block delineation and feature extraction. *Can. J. Remote Sens.* **2003**, *29*, 441–452. [CrossRef]

30. Belgiu, M.; Drăgut, L. Random forest in remote sensing: A review of applications and future directions. *ISPRS J. Photogramm. Remote Sens.* **2016**, *114*, 24–31. [CrossRef]

31. Clinton, N.; Holt, A.; Scarborough, J.; Yan, L.; Gong, P. Accuracy assessment measures for object-based image segmentation goodness. *Photogramm. Eng. Remote Sens.* **2010**, *76*, 289–299. [CrossRef]

32. Costa, H.; Foody, G.M.; Boyd, D.S. Supervised methods of image segmentation accuracy assessment in land cover mapping. *Remote Sens. Environ.* **2018**, *205*, 338–351. [CrossRef]

33. Belgiu, M.; Drăgut, L. Comparing supervised and unsupervised multiresolution segmentation approaches for extracting buildings from very high resolution imagery. *ISPRS J. Photogramm. Remote Sens.* **2014**, *96*, 67–75. [CrossRef] [PubMed]

34. Drăgut, L.; Tiede, D.; Levick, S.R. ESP: A tool to estimate scale parameter for multiresolution image segmentation of remotely sensed data. *Int. J. Geogr. Inf. Sci.* **2010**, *24*, 859–871. [CrossRef]

35. Kavzoglu, T.; Erdemir, M.Y.; Tonbul, H. Classification of semiurban landscapes from very high-resolution satellite images using a regionalized multiscale segmentation approach. *J. Appl. Remote Sens.* **2017**, *11*, 35016. [CrossRef]

36. Johnson, B.; Xie, Z. Unsupervised image segmentation evaluation and refinement using a multi-scale approach. *ISPRS J. Photogramm. Remote Sens.* **2011**, *66*, 473–483. [CrossRef]

37. Dragut, L.; Csillik, O.; Eisank, C.; Tiede, D. Automated parameterisation for multi-scale image segmentation on multiple layers. *ISPRS J. Photogramm. Remote Sens.* **2014**, *88*, 119–127. [CrossRef] [PubMed]

38. Espindola, G.M.; Camara, G.; Reis, I.A.; Bins, L.S.; Monteiro, A.M. Parameter selection for region-growing image segmentation algorithms using spatial autocorrelation. *Int. J. Remote Sens.* **2006**, *27*, 3035–3040. [CrossRef]

39. Zhang, X.; Feng, X.; Xiao, P.; He, G.; Zhu, L. Segmentation quality evaluation using region-based precision and recall measures for remote sensing images. *ISPRS J. Photogramm. Remote Sens.* **2015**, *102*, 73–84. [CrossRef]

40. Grippa, T.; Lennert, M.; Beaumont, B.; Vanhuysse, S.; Stephenne, N.; Wolff, E. An open-source semi-automated processing chain for urban obia classification. In Proceedings of the GEOBIA 2016: Solutions and Synergies, Enschede, The Netherlands, 14–16 September 2016.

41. Li, M.; Ma, L.; Blaschke, T.; Cheng, L.; Tiede, D. A systematic comparison of different object-based classification techniques using high spatial resolution imagery in agricultural environments. *Int. J. Appl. Earth Obs. Geoinf.* **2016**, *49*, 87–98. [CrossRef]

42. Ma, L.; Li, M.; Ma, X.; Cheng, L.; Du, P.; Liu, Y. A review of supervised object-based land-cover image classification. *ISPRS J. Photogramm. Remote Sens.* **2017**, *130*, 277–293. [CrossRef]

43. Cánovas-García, F.; Alonso-Sarría, F. A local approach to optimize the scale parameter in multiresolution segmentation for multispectral imagery. *Geocarto Int.* **2015**, *30*, 937–961. [CrossRef]

44. Grippa, T.; Georganos, S.; Vanhuysse, S.G.; Lennert, M.; Wolff, E. A local segmentation parameter optimization approach for mapping heterogeneous urban environments using VHR imagery. *Remote Sens. Technol. Appl. Urban Environ. II* **2017**, *10431*. [CrossRef]

45. Gorelick, N.; Hancher, M.; Dixon, M.; Ilyushchenko, S.; Thau, D.; Moore, R. Remote sensing of environment google earth engine: Planetary-scale geospatial analysis for everyone. *Remote Sens. Environ.* **2017**, *202*, 18–27. [CrossRef]

46. Tobler, W.R. A computer movie simulating urban growth in the detroit region. *Econ. Geogr.* **1970**, *46*, 234–240. [CrossRef]

47. Neteler, M.; Bowman, M.H.; Landa, M.; Metz, M. GRASS GIS: A multi-purpose open source GIS. *Environ. Model. Softw.* **2012**, *31*, 124–130. [CrossRef]

48. Grippa, T.; Georganos, S.; Zarougui, S.; Bognounou, P.; Diboulo, E.; Forget, Y.; Lennert, M.; Vanhuysse, S.; Mboga, N.; Wolff, E. Mapping urban land use at street block level using open street map, Remote Sensing Data, and Spatial Metrics. *ISPRS Int. J. Geo-Inform.* **2018**, *7*, 246. [CrossRef]

49. United Nations. *World Urbanization Prospects: The 2014 Revision, Highlights*, 3rd ed.; Population Division, United Nations: New York, NY, USA, 2014.

50. Schug, F.; Okujeni, A.; Hauer, J.; Hostert, P.; Nielsen Jonas Øand van der Linden, S. Mapping patterns of urban development in Ouagadougou, Burkina Faso, using machine learning regression modeling with bi-seasonal Landsat time series. *Remote Sens. Environ.* **2018**, *210*, 217–228. [CrossRef]

51. Momsen, E.; Metz, M.; GRASS Development TEAM. Module i.segment 2015. Available online: https://grass.osgeo.org/grass75/manuals/i.segment.html (accessed on 1 August 2018).

52. Böck, S.; Immitzer, M.; Atzberger, C. On the objectivity of the objective function—Problems with unsupervised segmentation evaluation based on global score and a possible remedy. *Remote Sens.* **2017**, *9*, 769. [CrossRef]

53. Georganos, S.; Lennert, M.; Grippa, T.; Vanhuysse, S.; Johnson, B.; Wolff, E. Normalization in unsupervised segmentation parameter optimization: A solution based on local regression trend analysis. *Remote Sens.* **2018**, *10*, 222. [CrossRef]

54. Georganos, S.; Grippa, T.; Lennert, M.; Vanhuysse, S.G.; Wolff, E. SPUSPO: Spatially Partitioned Unsupervised Segmentation Parameter Optimization for Efficiently Segmenting Large Heterogeneous Areas. In Proceedings of the 2017 Conference on Big Data from Space (BiDS'17), Toulouse, France, 28–30 November 2017.

55. Lennert, M.; GRASS Development TEAM. Module i.segment.uspo 2017. Available online: https://grass.osgeo.org/grass74/manuals/addons/i.segment.uspo.html (accessed on 1 August 2018).

56. Körting, T.S.; Castejon, E.F.; Fonseca, L.M.G. The divide and segment method for parallel image segmentation. In Proceedings of the International Conference on Advanced Concepts for Intelligent Vision Systems, Antwerp, Belgium, 18–21 September 2013.

57. Soares, A.R.; Körting, T.S.; Fonseca, L.M.G. Improvements of the divide and segment method for parallel image segmentation. *Rev. Bras. Cartogr.* **2016**, *68*.

58. Satnik, D.; GRASS Development TEAM. Module i.zc 2016. Available online: https://grass.osgeo.org/grass70/manuals/i.zc.html (accessed on 1 August 2018).

59. Lennert, M.; GRASS Development TEAM. Module i.cutlines 2018. Available online: https://grass.osgeo.org/grass74/manuals/addons/i.cutlines.html (accessed on 1 August 2018).

60. Osborne, P.E.; Foody, G.M.; Suárez-Seoane, S. Non-stationarity and local approaches to modelling the distributions of wildlife. *Divers. Distrib.* **2007**, *13*, 313–323. [CrossRef]

61. Chen, T.; Guestrin, C. XGBoost: Reliable large-scale tree boosting system. *arXiv* **2016**. [CrossRef]

62. Xia, Y.; Liu, C.; Li, Y.; Liu, N. A boosted decision tree approach using Bayesian hyper-parameter optimization for credit scoring. *Expert Syst. Appl.* **2017**, *78*, 225–241. [CrossRef]

63. Genuer, R.; Poggi, J.M.; Tuleau-Malot, C. VSURF: An R Package for variable selection using random forests. *R J.* **2015**, *7*, 19–33.

64. Georganos, S.; Grippa, T.; Vanhuysse, S.; Lennert, M.; Shimoni, M.; Kalogirou, S.; Wolff, E. Less is more: Optimizing classification performance through feature selection in a very-high-resolution remote sensing object-based urban application. *GISci. Remote Sens.* **2017**, 221–242. [CrossRef]

65. Georganos, S.; Grippa, T.; Lennert, M.; Johnson, B.A.; Vanhuysse, S.; Wolff, E. SPUSPO: Spatially Partitioned Unsupervised Segmentation Parameter Optimization for Efficiently Segmenting Large Heterogeneous Areas. Available online: https://zenodo.org/record/1341116#.W5S1oVKtZS0 (accessed on 31 August 2018).

66. Woodcock, C.E.; Strahler, A.H. The factor of scale in remote sensing. *Remote Sens. Environ.* **1987**, *21*, 311–332. [CrossRef]

67. Fotheringham, A.S.; Brunsdon, C.; Charlton, M. Geographically weighted regression: The analysis of spatially varying relationships. *Am. J. Agric. Econom.* **2004**, *86*, 554–556. [CrossRef]

68. Liu, T.; Abd-elrahman, A.; Jon, M.; Wilhelm, V.L.; Liu, T.; Abd-elrahman, A.; Jon, M.; Wilhelm, V.L. Comparing fully convolutional networks, random forest, support vector machine, and patch-based deep convolutional neural networks for object-based wetland mapping using images from small unmanned aircraft system. *GISci. Remote Sens.* **2018**, *55*, 243–264. [CrossRef]

69. Liu, T.; Abd-Elrahman, A. Deep convolutional neural network training enrichment using multi-view object-based analysis of Unmanned Aerial systems imagery for wetlands classification. *ISPRS J. Photogramm. Remote Sens.* **2018**, *139*, 154–170. [CrossRef]

70. Marmanis, D.; Schindler, K.; Wegner, J.D.; Galliani, S.; Datcu, M.; Stilla, U. Classification with an edge: Improving semantic image segmentation with boundary detection. *ISPRS J. Photogramm. Remote Sens.* **2018**, *135*, 158–172. [CrossRef]

71. Linard, C.; Tatem, A.J.; Gilbert, M. Modelling spatial patterns of urban growth in Africa. *Appl. Geogr.* **2013**, *44*, 23–32. [CrossRef] [PubMed]

72. Sandborn, A.; Engstrom, R.N. Determining the Relationship between Census Data and Spatial Features Derived From High-Resolution Imagery in Accra, Ghana. *IEEE J. Sel. Top. Appl. Earth Obs. Remote Sens.* **2016**, *9*, 1970–1977. [CrossRef]

73. Ming, D.; Li, J.; Wang, J.; Zhang, M. Scale parameter selection by spatial statistics for GeOBIA: Using mean-shift based multi-scale segmentation as an example. *ISPRS J. Photogramm. Remote Sens.* **2015**, *106*, 28–41. [CrossRef]

74. Yuan, Q.; Zhang, L.; Shen, H. Hyperspectral image denoising employing a spectral-spatial adaptive total variation model. *IEEE Trans. Geosci. Remote Sens.* **2012**, *50*, 3660–3677. [CrossRef]

75. Gu, H.; Han, Y.; Yang, Y.; Li, H.; Liu, Z.; Soergel, U.; Blaschke, T.; Cui, S. An efficient parallel multi-scale segmentation method for remote sensing imagery. *Remote Sens.* **2018**, *10*, 590. [CrossRef]

Article

Multiscale Optimized Segmentation of Urban Green Cover in High Resolution Remote Sensing Image

Pengfeng Xiao [1,2,3,*], **Xueliang Zhang** [1,2,3,*], **Hongmin Zhang** [1], **Rui Hu** [1] and **Xuezhi Feng** [1,2,3]

[1] Department of Geographic Information Science, School of Geography and Ocean Science,
    Nanjing University, Nanjing 210023, Jiangsu, China; zhm1054549500@gmail.com (H.Z.);
    njuhurui@163.com (R.H.); xzf@nju.edu.cn (X.F.)
[2] Collaborative Innovation Center of South China Sea Studies, Nanjing 210023, Jiangsu, China
[3] Jiangsu Center for Collaborative Innovation in Geographical Information Resource Development and
    Application, Nanjing 210023, Jiangsu, China
*   Correspondence: xiaopf@nju.edu.cn (P.X.); zxl@nju.edu.cn (X.Z.); Tel.: +86-25-8968-0612 (P.X. & X.Z.)

Received: 22 October 2018; Accepted: 13 November 2018; Published: 15 November 2018

Abstract: The urban green cover in high-spatial resolution (HR) remote sensing images have obvious multiscale characteristics, it is thus not possible to properly segment all features using a single segmentation scale because over-segmentation or under-segmentation often occurs. In this study, an unsupervised cross-scale optimization method specifically for urban green cover segmentation is proposed. A global optimal segmentation is first selected from multiscale segmentation results by using an optimization indicator. The regions in the global optimal segmentation are then isolated into under- and fine-segmentation parts. The under-segmentation regions are further locally refined by using the same indicator as that in global optimization. Finally, the fine-segmentation part and the refined under-segmentation part are combined to obtain the final cross-scale optimized result. The green cover objects can be segmented at their specific optimal segmentation scales in the optimized segmentation result to reduce both under- and over-segmentation errors. Experimental results on two test HR datasets verify the effectiveness of the proposed method.

Keywords: multiscale segmentation; scale parameter; cross-scale optimization; segmentation refinement; urban green cover

1. Introduction

Urban green cover can be defined as the layer of leaves, branches, and stems of trees and shrubs and the leaves of grasses that cover the urban ground when viewed from above [1]. This term is typically used to refer to urban green space identified from remote sensing data, as it is in this study. Green space is an essential infrastructure in cities because it provides various products and ecosystem services for urban dwellers that can address support to climate-change mitigation and adaptation, human health and well-being, biodiversity conservation, and disaster risk reduction [2]. Therefore, inventorying the spatial distribution of urban green cover is imperative in decision-making about urban management and planning [3].

High-spatial resolution (HR) remote sensing data have shown great potential in identifying both the extent and the corresponding attributes of urban green cover [4–8]. In order to fully exploit the information content of the HR images, geographic object-based image analysis (GEOBIA) has become the principal method [9] and has been successfully applied for urban green cover extraction [10–15]. Scale is a crucial aspect in GEOBIA as it describes the magnitude or the level of aggregation and abstraction on which a certain phenomenon can be described [16]. GEOBIA is sensitive to segmentation scale but has challenges in selecting scale parameters, because different objects can only be perfectly expressed at the scale corresponding to their own granularity. The urban green cover in HR images presents obvious multiscale characteristics, for example, the size of urban green cover varies in a large extent of scales; it can either be a small area with several square meters, such as a private garden, or reach a large area with several square kilometers such as a park. As a result, it is be possible to properly segment all features in a scene using a single segmentation scale, resulting in that over-segmentation (producing too many segments) or under-segmentation (producing too few segments) often occurs [17]. Therefore, it plays a decisive role in GEOBIA that divide the complex features at the appropriate scale to segment landscape into non-overlapping homogenous regions [18].

In order to find the optimal scale for each object, the multiscale segmentation can be optimized using three different strategies based on: supervised evaluation measures, unsupervised evaluation measures, and cross-scale optimization. (1) The supervised strategy compares segmentation results with reference by geometric [19–23] and arithmetic [21,24,25] discrepancy. This strategy is apparently effective but is, in fact, subjective and time-consuming when creating the reference. (2) The unsupervised strategy defines quality measures, such as intra-region spectral homogeneity [26–31] and inter-region spectral heterogeneity [32–34], for conditions to be satisfied by an ideal segmentation. It thus characterizes segmentation algorithms by computing goodness measures based on segmentation results without the reference. This strategy is objective but has the added difficulty of designing effective measures. (3) The cross-scale strategy fuses multiscale segmentations to achieve the expression of various granularity of objects at their optimal scale [35–37]. It can make better use of the multiscale information than the other two strategies.

Recently, cross-scale strategy has garnered much attention in the multiscale segmentation optimization by using evaluation measures as the indicator. (1) For the unsupervised indicator, some studies generated a single optimal segmentation by fusing multiscale segmentations according to local-oriented unsupervised evaluation [35,38,39]. However, the range of involved scales was found to be limited. By contrast, multiple segmentation scales were selected according to a change in homogeneity [27–29]. (2) For the supervised indicator, multiscale segmentation optimization has been achieved by using the single-scale evaluation measure based on different sets of reference objects [28]. For example, some studies have provided reference objects and suitable segmentation scales for different land cover types [40,41]. The difficulty of this strategy is preparing appropriate sets of reference objects that can reflect changes of scales. In our previous work [37], two discrepancy measures are proposed to assess multiscale segmentation accuracy: the multiscale object accuracy (MOA) measure at object level and the bidirectional consistency accuracy (BCA) measure at pixel level. The evaluation results show that the proposed measures can assess multiscale segmentation accuracy and indicate the manner in which multiple segmentation scales can be selected. These proposed measures can manage various combinations of multiple segmentation scales. Therefore, applications for optimization of multiscale segmentation can be expanded.

In this study, an unsupervised cross-scale optimization method specifically for urban green cover segmentation is proposed. A global optimal segmentation is first selected from multiscale segmentation results by using an optimization indicator. The regions in the global optimal segmentation are then isolated into under- and fine-segmentation parts. The under-segmentation regions are further locally refined by using the same indicator as that in global optimization. Finally, the fine-segmentation part and the refined under-segmented part are combined to obtain the final cross-scale optimization result. The goal of the proposed method is to segment urban vegetation in general, for example, trees and grass together included in one region. The segmentation result of urban green cover can be practically used in urban planning, for example investigation of urban green cover rate [42,43], and urban environment monitoring, for example influence analysis of the urban green cover to residential quality [44,45].

The contribution of this study is to propose a new cross-scale optimization method specifically for urban green cover to achieve the optimal segmentation scale for each green cover object. The same optimization indicator is designed to be used both to identify the global optimal scale and to refine the under-segmentation. By refining the isolated under-segmented regions for urban green cover, the optimization result can avoid under-segmentation errors as well as reduce over-segmentation errors, achieving higher segmentation accuracies than single-scale segmentation results. The proposed method also holds the potentials to be applied to cross-scale segmentation optimization for different types of urban green cover or even other land cover types by designing proper under-segmentation isolation rule.

The rest of the paper is organized as follows. Section 2 presents the proposed method of multiscale segmentation optimization. Section 3 describes the study area and test data. Section 4 verifies the effectiveness of the proposed method based on experiments. Section 5 presents the discussions. Finally, conclusions are drawn in Section 6.

2. Method

2.1. General Framework

This study proposes a multiscale optimization method for urban green cover segmentation, which aims to comprehensively utilize multiscale segmentation results to achieve optimal scale expression of urban green cover. Figure 1 shows the general framework of the proposed method. First, a global optimal segmentation is selected from the multiscale segmentation results by using an optimization indicator. The indicator is the local peak (*LP*) of the change rate (*CR*) of the mean value of spectral standard deviation (*SD*) of each segment. Second, the regions in the global optimal segmentation result are isolated into under- and fine-segmentation parts, based on designed under-segmentation isolation rule. Third, the under-segmentation regions are refined by using the same optimization indicator *LP* in a local version. Finally, the fine-segmented part and the optimized under-segmented part are combined to obtain the final cross-scale optimization result.

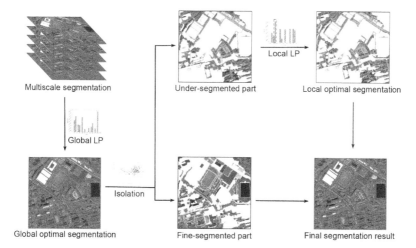

Figure 1. General framework of the proposed multiscale segmentation optimization method. The remote sensing images are shown with false color composite: red: near infrared band; green: red band; and blue: green band. The multiscale, under-segmented, and fine-segmented regions are shown with blue, green, and yellow polygons.

2.2. Hierarchical Multiscale Segmentation

The hierarchical multiscale segmentation is composed of multiple segments from fine to coarse at each location, in which the small objects are supposed to be represented by fine segments at certain segmentation scales and the large objects are represented by coarse segments correspondingly. Furthermore, a fine-scale segment smaller than a real object is supposed to represent a part of the object, while a coarse-scale segment larger than a real object is to represent an object group. A preliminary requirement for the multiscale segments is that the segments at the same location should be nested. Otherwise the object boundaries would be conflict when combining or fusing the multiscale segments.

The hierarchical multiscale segmentation is represented using a segment tree model [46], as shown in Figure 2. The tree nodes at different levels represent segments at different scales. An arc connecting a parent and a child node represents the inclusion relation between segments at adjacent scales. The leaf nodes represent the segments at the finest scale and the nodes at upper levels represent segments at coarser scales. Finally, the root node represents the whole image. An ancestry path in the tree is defined as the path from a leaf node up to the root node, revealing the transition from object part to the whole scene. The hierarchical context of each leaf node is conveyed by the ancestry path, in which a segment is gradually becoming coarser and finally reaching the whole image.

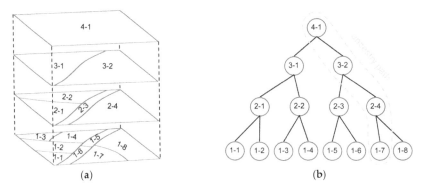

(a)　　　　　　　　　　　　　　(b)

Figure 2. Illustration of segment tree model (**b**) that represents hierarchical multiscale segmentation (**a**).

Several region-based segmentation methods can be applied to produce the required hierarchical multiscale segmentations, for example multiresolution segmentation method [47], mean-shift method [48] and hierarchical method [49]. Specifically, the multiresolution segmentation method [47] is used in the study, in which the shape parameter is set as 0.5 by default. The regions at each segmentation scale are represented by the nodes at the same level in the segment tree. Finally, the segment tree is constructed by recording the multiscale segmentation.

2.3. Selecting Global Optimal Scale

We need to first select a global optimal segmentation scale and then refine the under-segmentation part for urban green cover. Thus, unlike other optimal scale selection methods for compromising under- and over-segmentation errors, we design an indicator to select an optimal scale in which segmentation results mainly include reasonable under-segmented and fine-segmented regions, reducing over-segmented regions as much as possible.

Referring to the standard deviation indicator [28], we adopt the indicator focusing on homogeneity of segments by calculating the mean value of spectral standard deviation (*SD*) of each segment. *SD* is defined as below:

$$SD = \sqrt{\frac{1}{nb} \sum_{k=1}^{b} \sum_{i=1}^{n} SD_{ki}} \qquad (1)$$

where SD_{ki} is the standard deviation of digital number (DN) of spectral band k in segment i; n is the number of segments in the image; and b is the number of spectral bands of the image.

With the increase of the scale parameter, *SD* will change as following. Generally, it tends to increase because the homogeneity of segments is gradually decreased in the region merging procedure.

Near the scale that the segments are close to the real objects, the change rate of *SD* will increase suddenly because of the influence of the boundary pixels [29].

To find the scale in which the green cover segments are closest to the real objects, we propose indicator *CR* to represent the change rate of *SD* and indicator *LP* to represent the local peak of *CR*. They are defined respectively as below:

$$CR = \frac{dSD}{dl} = \frac{SD(l) - SD(l - \Delta l)}{\Delta l} \tag{2}$$

$$LP = [CR(l) - CR(l - \Delta l)] + [CR(l) - CR(l + \Delta l)] \tag{3}$$

where *l* is the segmentation scale and Δl is the increment in scale parameter, that is the lag at which the scale parameter grows. The scale increment has powerful control over the global optimal segmentation because it can smooth the heterogeneity measure resulting in the optimal segmentation occurred in different scales [50]. Experimentally, the small increments (e.g., 1) produces optimal segmentation in finer scales while the large increments (e.g., 100) produces optimal segmentation in coarser scales [28]. Hence, the medium increment (e.g., 10) of scale is adopted in the study.

According to the aforementioned change law of *SD*, near the scale that the segments are close to the real objects, the *CR* will increase suddenly because of the influence of the boundary pixels of green cover segments. Thus, a *LP* will appear when the global optimal segmentation scale is coming for several segments. However, there are several *LP*s within a set of increased scale parameters because not all the segments have the same optimal segmentation scale. The global optimal segmentation is identified as the scale with largest *LP*, because the largest *LP* indicates that most of the segments in the image reach the optimal segmentation state. Furthermore, the large *LP* could also be caused by the large *SD* value of coarse segments, because the large *SD* will produce large *CR* and corresponding large *LP*, revealing the under-segmentation state. Therefore, the next step is to optimize the under-segmentation part of the global optimal segmentation result for green cover objects.

2.4. Isolating Under-Segmented Regions

In order to obtain the under-segmented regions for green cover from the globally optimized segmentation result, further isolation of segments is required. When a green cover object is in the under-segmentation state, it is often mixed with other adjacent objects and the spectral standard deviation (SD_i) is thus great. Moreover, since the normalized difference vegetation index (NDVI) has a good performance to distinguish between green cover and other features, when other objects are mixed with the green cover object, the NDVI value of the region is not very high, that is lower than that of green cover objects, as well as not very low, that is higher than that of non-green cover objects. NDVI is defined as the ratio of difference between near infrared band and red band values to their sum [51]. Thus, NDVI of a region is calculated as below:

$$NDVI_i = \frac{1}{m} \sum_{j=1}^{m} \frac{NIR_j - R_j}{NIR_j + R_j} \tag{4}$$

where NIR_j and R_j are DNs of near infrared and red band for pixel *j*, respectively; and *m* is the number of pixels in region *i*.

Therefore, a region with a high SD_i value and a medium $NDVI_i$ value can be considered an under-segmentation region for green cover. The isolation rule for an under-segmentation region with green cover are thus defined as below:

$$\begin{cases} SD_i > T_{SD} \\ T_{N1} < NDVI_i < T_{N2} \end{cases} \tag{5}$$

where T_{SD}, T_{N1}, and T_{N2} are thresholds that need to be set by users. We set it by the trial-and-error strategy. A segment with SD_i lower than T_{SD} is viewed as fine segment because of the high homogeneity. If the $NDVI_i$ value of region i is higher than T_{N2}, it is viewed as fine segmentation of green cover; and if it is lower than T_{N1}, it is viewed as not containing green cover and will not be involved in the successive refining procedure.

2.5. Refining Under-Segmented Regions

For each individual region in the under-segmentation part, the segment tree is first used to quantify the spatial context relationship of the regions at different scales and the appropriate segmentation scale is then selected through the optimization indicator *LP*. Finally, the under-segmentation part is replaced by the optimized segments. The specific steps are performed as follows:

(1) Select one under-segmented region R_i, extract the segmentations at lower scales than the global optimal scale in region R_i.
(2) Compute the *LP* of each scale and the local optimal scale of green cover is defined as scale with a largest *LP* in region R_i.
(3) Replace R_i with the local optimal scale segmentation.
(4) Repeat step (1)–(3) until all under-segmented regions are refined according to Equation (5).

2.6. Accuracy Assessment

Segmentation quality evaluation strategies include visual analysis, system-level evaluation, empirical goodness, and empirical discrepancy methods [37]. The last two methods are also known as unsupervised and supervised evaluation methods, respectively. The unsupervised evaluation method calculates indexes of homogeneity within segments and heterogeneity between segments [35]. It does not require ground truth but the explanatory of designing measures and the meaning of measure values is insufficient. The supervised evaluation method compares segmentation results with ground truth and its discrepancy can directly reveal the segmentation quality [52]. Region-based *precision* and *recall* measures are sensitive to both geometric and arithmetic errors. Thus, the supervised evaluation method is used to assess the segmentation accuracy of the multiscale optimization.

Precision is the ratio of true positives to the sum of true positives and false positives, and *recall* is the ratio of true positives to the sum of true positives and false negatives. Given the segmentation result S with n segments $\{S_1, S_2, \ldots, S_n\}$ and the reference R with m objects $\{R_1, R_2, \ldots, R_m\}$, the *precision* measure is calculated by matching $\{R_i\}$ to each segment S_i and the *recall* measure by matching $\{S_i\}$ to each reference object R_i. When calculating the *precision* measure, the matched reference object (R_{imax}) for each segment S_i is first identified, where R_{imax} has the largest overlapping area with S_i. The *precision* measure is then defined as [23]:

$$precision = \frac{\sum_{i=1}^{n}|S_i \cap R_{imax}|}{\sum_{i=1}^{n}|S_i|} \qquad (6)$$

where $|\cdot|$ denotes the area that is represented by the number of pixels in a region.

Similarly, the matched segment (S_{imax}) for each reference object R_i is searched according to the maximal overlapping area criterion and the *recall* measure is defined as [23]:

$$recall = \frac{\sum_{i=1}^{m}|R_i \cap S_{imax}|}{\sum_{i=1}^{m}|R_i|} \qquad (7)$$

The *precision* and *recall* measures both range from 0 to 1. Using these two measures can determine both under- and over-segmented situations. An under-segmented result will have a large *recall* and a low *precision*. By contrast, if the result is over-segmented, the *precision* is high but the *recall* is low. If the *precision* and *recall* values of one segmentation result are both higher than another, this result is considered to have a better segmentation quality. However, we do not know which one is better when one measure in larger than another and the other measure is smaller than another. Hence, we should combine these two measures into one. In this study, we use the harmonic average of *precision* and *recall* called *F-score* [53], which is defined as:

$$F - score = \frac{2 \cdot precsion \cdot recall}{precsion + recall} \qquad (8)$$

where an *F-score* reaches its best value at 1 (perfect *precision* and *recall*) and worst at 0.

3. Data

The study area is located in Nanjing City (32°02′38″N, 118°46′43″E), which is the capital of Jiangsu Province of China and the second largest city in the East China region (Figure 3), with an administrative area of 6587 km² and a total population of 8335 thousand as of 2017. As one of the four garden cities in China, Nanjing has a wealth of urban green spaces than many of other cities. The urban green cover rate in the built-up area of Nanjing is 44.85% in 2018.

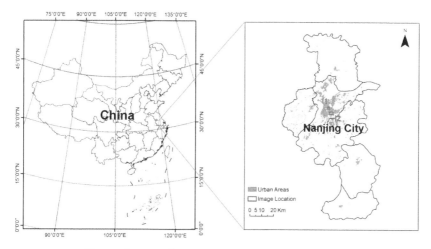

Figure 3. Location of the Nanjing City and the two test images.

In this study, an IKONOS-2 image acquired on 19 October 2010 and a WorldView-2 image acquired on 29 July 2015 in Nanjing are used as the HR data. Both the images consist of four spectral bands: blue, green, red, and near infrared. The spatial resolution of the multispectral bands of the IKONOS-2 image is improved from 3.2 m to 0.8 m after pan-sharpening. The spatial resolution of the multispectral bands of the WorldView-2 image is 2 m.

Two test images identified as I1 and I2 are subsets of the IKONOS-2 and the WorldView-2 images, respectively, containing urban green cover in traffic area, residential area, campus area, park area, commercial area, and industrial area, which are the typical areas in urban. The size of I1 and I2 are 2286 × 1880 and 1478 × 974 pixels and the area are approximate 2.8 km² and 5.8 km², respectively. As shown in Figure 4, there are abundant green cover objects distributed in the images and various in size and shape.

In order to evaluate the segmentation accuracy, we randomly select some green cover objects as reference. The reference objects are uniformly distributed in the test images and various in size and shape. Each reference object is delineated by one person and reviewed by other to catch any obvious errors. Finally, we collect 130 reference objects for each test image. It is noted that if there are trees covered a road, this area will be digitized as green cover objects. The area of the smallest reference object is only 59.5 m^2, whereas the area of the biggest reference object is 14,063.1 m^2. Hence, it is not possible to properly segment all of the green cover objects using a single segmentation scale.

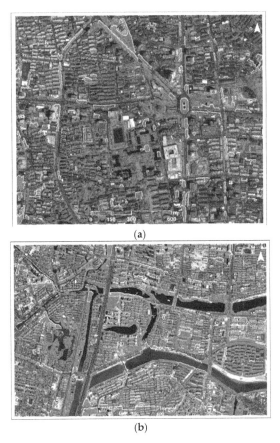

(a)

(b)

Figure 4. Test image I1 from an IKONOS-2 image (**a**) and test image I2 from a WorldView-2 image (**b**), containing different green cover in urban area. The images are shown with false color composite: red: near infrared band; green: red band; and blue: green band. The reference green cover objects are shown with orange polygons.

4. Results

4.1. Global Optimal Scale Selection

The multiscale segmentation results are produced by applying multiresolution segmentation method. For I1, the scale parameters are set from 10 to 250 by increment of 10. Since the spatial resolution of I2 is coarser than I1, the scale parameters are set from 10 to 125 by increment of 5. If we set the same scale parameters for I2 as those for I1, the coarse segmentation scales (e.g., >130) would be seriously under-segmented and the homogeneity of segments at these coarse scales would change randomly, which could not benefit the optimization procedure and could even do harm to the optimization procedure.

The multiscale segmentations cover apparently over-segmentation, medium segmentation, and apparently under-segmentation. The optimization indicators SD, CR, and LP are respectively calculated for each segmentation result and shown in Figure 5. When the scale parameter increases,

SD gradually increases, which indicates that the regions are gradually growing and the homogeneity decreases. Correspondingly, in the process of *SD* change, *CR* appears multiple local peaks. The indicator *LP* can highlight these local peaks of *CR* very well. We can see that *LP* appears at segmentation scales of 80, 110, 150, 170, 190, and 220 for I1, in which *LP* reaches the maximum at 220. For I2, *LP* appears at segmentation scales of 45, 55, 60, 70, 80, 85, 95, 105, and 120, where *LP* is the largest at 105. Therefore, the segmentation with the scale parameter of 220 and 105 is taken as the global optimal segmentation scale for I1 and I2.

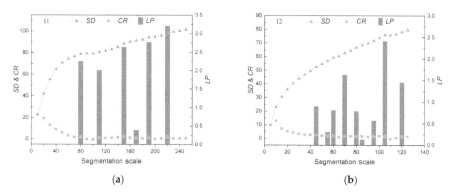

Figure 5. Changes of optimization indicator *SD*, *CR*, and *LP* with scale parameter for the multiscale segmentations of test image I1 (**a**) and I2 (**b**).

Combining with the supervised evaluation results of the multiscale segmentations (Figure 6), we can know that the selected global optimal segmentation scale is at the under-segmentation status for green cover objects. For both I1 and I2, the *precision* value is apparently lower than the *recall* value in the optimal scale, which indicates the under-segmentation status. To further illustrate this, the selected I2 segmentation result at scale 105 is presented in Figure 7, in which we can clearly see that except for several fine-segmented green cover objects with relatively large size, many green cover objects are shown as under-segmented.

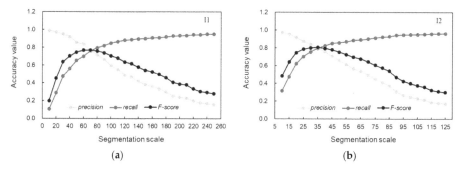

Figure 6. Changes of *precision*, *recall*, and *F-score* with scale parameter for the multiscale segmentation results of test image I1 (**a**) and I2 (**b**).

The selected global optimal segmentation of green cover tends to appear in the case of coarse scales. As a result, the over-segmentation errors are reduced, while some green cover objects with small size will inevitably be in an under-segmentation state and single scale cannot achieve optimal segmentation of green cover objects of different sizes. Therefore, it is necessary to further optimize the global optimal segmentation by refining the under-segmented regions.

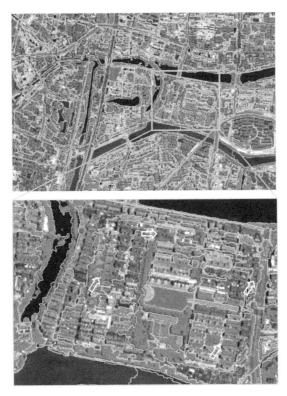

Figure 7. Selected global optimal segmentation at the scale of 105 for I2. The segments are shown with green polygons, the examples of fine-segmentation for green cover are shown with blue polygons, and those of under-segmentation for green cover are marked by yellow arrows.

4.2. Under-Segmented Region Isolation

The under-segmented regions are isolated by the rule in Equation (5). The threshold values of T_{SD}, T_{N1}, and T_{N2} are set as 40, 0.05, and 0.25 for test image I1 and as 40, 0.10, and 0.55 for test image I2. The threshold values of $NDVI_i$ for I2 is set as different for I1, this is mainly because the different acquisition date between I1 and I2, between which the vegetation growth status is different.

To illustrate the effectiveness of the designed isolation rule for under-segmentation with green cover, several sample segments in the global optimal segmentations are presented in Figure 8. It can be seen that the under-segmented regions containing green cover have medium $NDVI_i$ values and high SD_i values, as shown in Figure 8a,b,f,g. The fine-segmentation of green cover present high $NDVI_i$ values as shown in Figure 8c,i. A special case of fine-segmentation is shown in Figure 8h, which is a segment mainly containing sparse grass and the $NDVI_i$ value is thus not very high. However, the relatively low SD_i value of grass segment can prevent it to be wrongly identified as under-segmentation. The segments without green cover usually present low $NDVI_i$ value as shown in Figure 8d. A special case of segment without green cover is shown in Figure 8e, where the roof segment has a medium $NDVI_i$ value because of the roof material. However, the relatively low SD_i value can prevent it to be wrongly identified as under-segmentation containing green cover.

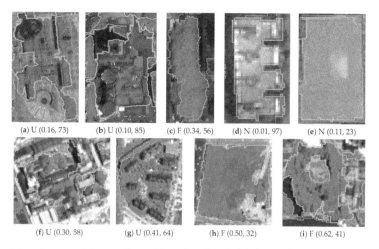

(**a**) U (0.16, 73)   (**b**) U (0.10, 85)   (**c**) F (0.34, 56)   (**d**) N (0.01, 97)   (**e**) N (0.11, 23)

(**f**) U (0.30, 58)   (**g**) U (0.41, 64)   (**h**) F (0.50, 32)   (**i**) F (0.62, 41)

Figure 8. Sample segments (cyan polygons) from the selected global optimal segmentation to illustrate the effectiveness of the designed isolation rule for under-segmentation containing green cover. (**a–e**) are from the results of test image I1 and (**f–i**) are from the results of test image I2. U, F, and N represent under-segmentation, fine-segmentation, and non-green-cover segmentation, respectively. The numbers in the bracket are sequentially the $NDVI_i$ and SD_i values of the segment.

To further validate the effectiveness of the isolation rule, the up-left part of the isolation results of I1 and I2 is zoomed in Figure 9. It can be seen that the green cover and other objects are mixed in the isolated under-segmented regions. In the fine-segmentation part, the regions are either fine-segmented green cover or segments without green cover.

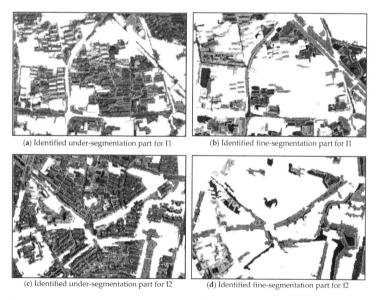

(**a**) Identified under-segmentation part for I1   (**b**) Identified fine-segmentation part for I1

(**c**) Identified under-segmentation part for I2   (**d**) Identified fine-segmentation part for I2

Figure 9. Isolation results of under-segmentation and fine-segmentation using the designed isolation rule. (**a,b**) are the up-left part of result of test image I1 and (**c,d**) are the up-left part of result of test image I2. The segments are shown with green polygons.

4.3. Under-Segmented Region Refinement

The multiscale optimized segmentation is obtained by refining the under-segmented part of the global optimal scale. In the refinement segmentation result, with the benefit of cross-scale refinement strategy, the segments are at different segmentation scales to achieve the optimal segmentation scale for each green cover object. The histogram of segmentation scales in the refinement results is shown in Figure 10. The refined segments almost cover all the segmentation scales finer than the selected global optimal scale. There are many segments at the small segmentation scales, for example scale 20 to 40 for I1 and scale 15 to 25 for I2, because there are many small-size green cover objects in urban area, such as single trees.

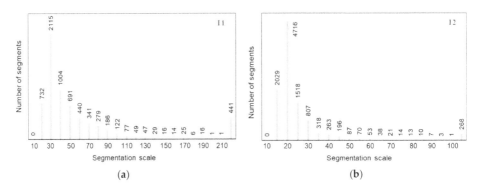

(a) (b)

Figure 10. Histogram of segmentation scales in the refinement results of test image I1 (**a**) and I2 (**b**).

To illustrate the effectiveness of achieving optimal scale for each green cover object, the sample refinement results from test images are enlarged to present in Figure 11 with labels of the scale number for each segment. Generally, it can be seen that large green cover objects are segmented at relatively coarse segmentation scales while small green cover objects are segmented at relatively small segmentation scales. The green cover objects, especially the small ones, tend to be segmented by a single segment.

(a) (b)

Figure 11. Sample refinement results of test image I1 (**a**) and I2 (**b**) labeled with optimized segmentation scale number. The segments are shown with green polygons.

The supervised evaluation results of segmentation before and after refinement are presented in Table 1 to quantify the effectiveness of the under-segmentation refinement. It can be seen that the *precision* value is apparently improved after refinement, showing that the under-segmented green cover objects can be effectively refined. The *recall* value is decreased mainly because the reduced under-segmentation. Therefore, the *F-score* after refinement is apparent improved than that before refinement. The segmentation results before and after refinement shown in Figure 12 further prove this.

To quantify the effectiveness of cross-scale optimization, the refinement result is compared with single-scale segmentation that has the highest *F-score* in the produced multiscale segmentations, which is at scale 70 for I1 and 35 for I2. The supervised evaluation results are also presented in Table 1. It can be seen that the *precision* of the refinement result is slightly lower than that of the single-scale best result while the *recall* is higher, which could be caused by the over-segmentation of the large green cover objects. Another reason for the lower *precision* for the refinement result could be caused by the wrong identification of under-segmented green cover objects, which makes the under-segmentation cannot be refined and thus lowers the precision accuracy. As a whole, the *F-score* of the refinement result is slightly higher than that of the single-scale best segmentation, which could mainly be caused by the reduced under-segmentation errors in the refining procedure. The segmentation results presented in Figure 11 further show the difference. As highlighted by the yellow rectangles, the existed under-segmentation errors in the single-scale best segmentation can be effectively reduced by the proposed refining strategy, which indicates the effectiveness of the refining procedure on overcoming under-segmentation errors. According to the comparison result with single-scale best segmentation, we can safely conclude that the proposed unsupervised multiscale optimization method can automatically produce optimal segmentation result at least equals to single-scale best segmentation indicated by supervised evaluation. Furthermore, the proposed refining strategy can help to reduce under-segmentation errors even in the single-scale best segmentation.

Table 1. Comparisons of the segmentation accuracy for the result before refinement, after refinement, and the single-scale best result.

	I1			I2		
	precision	*recall*	*F-score*	*precision*	*recall*	*F-score*
Before refinement	0.204	0.948	0.336	0.234	0.955	0.376
After refinement	0.764	0.811	0.787	0.766	0.859	0.810
Single-scale best	0.773	0.766	0.770	0.812	0.801	0.806

Figure 12. Comparison of sample segmentation results of test image I1 and I2 before refinement (the first row), after refinement (the second row), and the single-scale best segmentation result according to *F-score* (the third row), which are shown with yellow, green, and pink polygons, respectively. Four areas are highlighted for comparison using orange rectangles.

5. Discussions

5.1. Influence of Selected Global Optimal Segmentation Scale

The selected global segmentation scale is assumed to be reasonably under-segmented, which means that a part of segments is under-segmented and others are fine-segmented with few over-segmentation errors. This is because the successive refining procedure is designed to reduce the under-segmentation errors rather than over-segmentation errors. The results in Section 4.1 proved that the used optimization indicator can select segmentation scale at under-segmentation status. However, we can see from Figure 7 that many segments in the selected segmentation are extensively under-segmented, especially for those containing small green cover objects. Actually, the extensive under-segmentation could make the refinement difficult because too many different objects are mixed, which makes the segment features become erratic or random.

Even though the automatically selected global optimal segmentation scale can result in satisfactory refinement result, we explored whether a less under-segmentation than the selected one could further improve the refinement performance. For test images I1 and I2, a less under-segmented scale than the automatically selected one is input to the refining procedure and the segmentation accuracies of the refinement results are presented in Table 2. It is noted that the less under-segmented scale is also a local peak in Figure 5. We can see that the less under-segmented global segmentation scale could result in higher *precision* and higher *F-score* accuracies, which is because the prevalent under-segmentation is reduced that achieves better refinement performance. The sample segments of I1 shown in Figure 13 illustrate the difference. The segment at scale 220 in Figure 13a is prevalent under-segmented in terms of the green cover object, the $NDVI_i$ value of which is thus very low and it is not identified as the under-segmentation by the rule. Therefore, it is not involved in the following refinement procedure and the under-segmentation error is presented in the refinement result. However, the corresponded segment at scale 110 is less under-segmented and the $NDVI_i$ value is higher than that at scale 220, which make it allowed to be refined to remove the under-segmentation error. As a whole, this example demonstrates the importance of selecting a reasonable under-segmentation scale for the successive refinement. Specifically, the segments should better not be prevalent under-segmented.

Table 2. Comparisons of the segmentation accuracy for the refinement results from different global segmentation scale.

I1				I2			
Global Scale	*precision*	*recall*	*F-score*	Global Scale	*precision*	*recall*	*F-score*
220	0.764	0.811	0.787	105	0.766	0.859	0.810
110	0.826	0.788	0.807	70	0.791	0.846	0.817

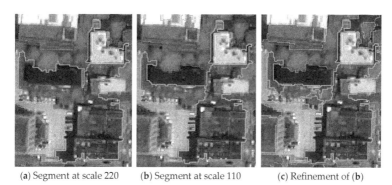

(a) Segment at scale 220   (b) Segment at scale 110   (c) Refinement of (b)

Figure 13. Sample segments (cyan polygons) of image I1 to illustrate the influence of global segmentation scale to refinement result. The $NDVI_i$ and SD_i values for the segment containing green cover object are 0.04 and 0.91 in (a) and 0.12 and 0.89 in (b) and (c) is the refinement result of (b).

5.2. Key Role of Identifying Under-Segmented Region

Based on the global segmentation scale with under-segmentation, the identification of under-segmented region serves as a key role to reduce under-segmentation errors once the refinement procedure is effective. Hence, before illustrating the key role of identifying under-segmented region, the effectiveness of refinement procedure is further judged in addition to Section 4.3.

Since the optimization indicator of maximum *LP* tends to select under-segmented scale, it makes the refining procedure performed iteratively because the selected optimal scale in the refining procedure could still be under-segmented for some segments and these segments need to be further refined. This increases the calculations for the refinement but it can lead to the safe refinement that avoids new over-segmentation errors.

The segmentation accuracies in the refining procedure of image I1 are presented in Table 3, through which we can see that the *F-score* is gradually increased. This is caused by the iteratively reduced under-segmentation errors in the refining procedure, which shows the effectiveness of the refining strategy and also the effectiveness of the isolation rule for under-segmented region. The sample segments in Figure 14 further demonstrates the effectiveness of the refining procedure.

As discussed above, the identification of under-segmented region serves as a key role in the proposed method. If the under-segmented regions can be correctly identified, the refinement can achieve success on removing under-segmentation errors because the refining procedure stops when the segment is not identified as under-segmented. Furthermore, if the under-segmented regions cannot be correctly identified, it cannot even be involved into the refining procedure and the under-segmentation error would be preserved in the refinement result. The proposed isolation rules of this study is still primary, which need to set appropriate thresholds by users. Even though it is proved to be effective, the automatic isolation of under-segmentation is still needed to be explored in the future.

Table 3. Segmentation accuracies in the refining procedure of I1.

	precision	*recall*	*F-score*
Before refinement	0.204	0.948	0.336
First iteration	0.492	0.899	0.636
Second iteration	0.693	0.835	0.757
Third iteration	0.764	0.811	0.787

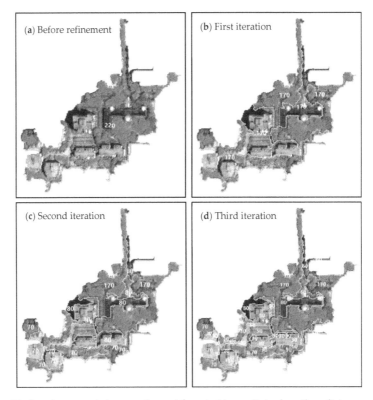

(a) Before refinement

(b) First iteration

(c) Second iteration

(d) Third iteration

Figure 14. Sample segments (green polygons) from test image I1 to show the refining procedure. The number of optimized segmentation scale is labeled in each segment.

5.3. Potential of Segmenting Different Types of Urban Green Cover

The proposed cross-scale optimization method aims at segment urban green cover into single regions by reducing both the under- and over-segmentation errors. The segmentation of urban green cover in general can be practically used in urban planning, for example investigation of urban green cover rate [42,43], and urban environment monitoring, for example influence analysis of the urban green cover to residential quality [44,45]. Surely, it would be more practical useful to segment different types of urban green cover [3,12]. To achieve this by the proposed method, the key step is to adjust isolation rule to identify the under-segmented regions containing different types of green cover. Since the optimization indicator used in the global segmentation scale selection and local refinement is based on the spectral standard deviation, it should be able to deal with under-segmentation for different green cover types. Once the under-segmentation for different types of green cover can be identified, the refining procedure is expected to be able to reduce the corresponding under-segmentation error and separate different types of green cover.

Actually, even though the presented method is not aiming at segmenting different types of green cover, several under-segmented segments containing different types of green cover are refined into each type when they meet the isolation rule in this study. An example of isolating different types of green cover based on scale 110 for image I1 is shown in Figure 15. This shows the potential of adjusting the under-segmentation isolating rule to achieve separating different types of green cover in the future.

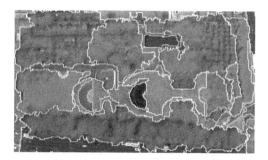

Figure 15. Sample segment from test image I1 to show the potential of discriminating different green cover types by the proposed method. The yellow polygons represent segments at scale 110 before refinement and the green polygons represent the new segments after refinement.

5.4. Potential of Refining Under-Segmented Regions for Other Land Cover Types

The under-segmented segments for urban green cover contain other land cover types, for example buildings, roads, and water. These segments are also under-segmented for the non-green-cover objects in it. When refining the under-segmentation for green cover, the under-segmentation errors for other land cover objects are also refined, as shown in Figures 11–15. That is to say, the non-green-cover objects in the identified under-segmented segments could also be refined along with the refinement of green cover objects. This reveals the potential of refining other land cover objects by applying the proposed optimization method.

As discussed above, the optimization indicator used in the global segmentation scale selection and local refinement should be able to deal with under-segmentation for different land cover types because it is based on the spectral standard deviation. Accordingly, if a proper isolation rule for other land cover types is designed, the proposed method is expected to be able to optimize the segmentations for those objects, which is also a direction of the future work based on this study.

6. Conclusions

In this paper, a multiscale optimized segmentation method for urban green cover is proposed. The global optimal segmentation result is first selected from the hierarchical multiscale segmentation results by using the optimization indicator global *LP*. Based on this, under-segmented regions and fine-segmented regions are isolated by the designed rule. For under-segmented regions, local *LP* is used for refinement, which ultimately allows urban green cover objects of different sizes to be expressed at their optimal scale.

The effectiveness of proposed cross-scale optimization method is proved by experiments based on two test HR images in Nanjing, China. With the benefit of cross-scale optimization, the proposed unsupervised multiscale optimization method can automatically produce optimal segmentation result with higher segmentation accuracy than single-scale best segmentation indicated by supervised evaluation. Furthermore, the proposed refining strategy is demonstrated to be able to effectively reduce under-segmentation errors.

The proposed method can be improved and extended in the future, for example optimizing segmentation for different types of urban green cover or even other land cover types. The key step is to design appropriate isolation rule of under-segmentation for specific applications. To further explore the potentials of the proposed method would be the main future work based on this study.

Author Contributions: Conceptualization, P.X. and X.Z.; Methodology, P.X., X.Z.; Software, H.Z., R.H.; Validation, H.Z. and X.Z.; Formal Analysis, H.Z. and X.Z.; Investigation, P.X.; Resources, P.X.; Data Curation, H.Z. and R.H.; Writing-Original Draft Preparation, P.X.; Writing-Review & Editing, P.X. and X.Z.; Supervision, P.X. and X.F.; Project Administration, P.X.; Funding Acquisition, P.X. and X.Z.

Funding: This research was funded by the National Natural Science Foundation of China grant number 41871235 and 41601366, the National Science and Technology Major Project of China grant number 21-Y20A06-9001-17/18, the Natural Science Foundation of Jiangsu Province grant number BK20160623, and the Open Research Fund of State Key Laboratory of Space-Ground Integrated Information Technology grant number 2016_SGIIT_KFJJ_YG_01.

Acknowledgments: The authors acknowledge the academic editors and anonymous reviewers for their insightful comments and suggestions helping to improve quality and acceptability of the manuscript.

Conflicts of Interest: The authors declare no conflict of interest.

References

1. Kanniah, K.D. Quantifying green cover change for sustainable urban planning: A case of Kuala Lumpur, Malaysia. *Urban For. Urban Green.* **2017**, *27*, 287–304. [CrossRef]
2. Salbitano, F.; Borelli, S.; Conigliaro, M.; Chen, Y. *Guidelines on Urban and Peri-Urban Forestry*; FAO Forestry Paper (FAO); FAO: Rome, Italy, 2016; ISBN 978-92-5-109442-6.
3. Wen, D.; Huang, X.; Liu, H.; Liao, W.; Zhang, L. Semantic Classification of Urban Trees Using Very High Resolution Satellite Imagery. *IEEE J. Sel. Top. Appl. Earth Obs. Remote Sens.* **2017**, *10*, 1413–1424. [CrossRef]
4. Nichol, J.; Lee, C.M. Urban vegetation monitoring in Hong Kong using high resolution multispectral images. *Int. J. Remote Sens.* **2005**, *26*, 903–918. [CrossRef]
5. Iovan, C.; Boldo, D.; Cord, M. Detection, Characterization, and Modeling Vegetation in Urban Areas from High-Resolution Aerial Imagery. *IEEE J. Sel. Top. Appl. Earth Obs. Remote Sens.* **2008**, *1*, 206–213. [CrossRef]
6. Ouma, Y.O.; Tateishi, R. Urban-trees extraction from Quickbird imagery using multiscale spectex-filtering and non-parametric classification. *ISPRS J. Photogramm. Remote Sens.* **2008**, *63*, 333–351. [CrossRef]
7. Tooke, T.R.; Coops, N.C.; Goodwin, N.R.; Voogt, J.A. Extracting urban vegetation characteristics using spectral mixture analysis and decision tree classifications. *Remote Sens. Environ.* **2009**, *113*, 398–407. [CrossRef]
8. Huang, X.; Lu, Q.; Zhang, L. A multi-index learning approach for classification of high-resolution remotely sensed images over urban areas. *ISPRS J. Photogramm. Remote Sens.* **2014**, *90*, 36–48. [CrossRef]
9. Blaschke, T.; Hay, G.J.; Kelly, M.; Lang, S.; Hofmann, P.; Addink, E.; Queiroz Feitosa, R.; van der Meer, F.; van der Werff, H.; van Coillie, F.; et al. Geographic Object-Based Image Analysis—Towards a new paradigm. *ISPRS J. Photogramm. Remote Sens.* **2014**, *87*, 180–191. [CrossRef] [PubMed]
10. Ardila, J.P.; Bijker, W.; Tolpekin, V.A.; Stein, A. Context-sensitive extraction of tree crown objects in urban areas using VHR satellite images. *Int. J. Appl. Earth Obs. Geoinf.* **2012**, *15*, 57–69. [CrossRef]

11. Yin, W.; Yang, J. Sub-pixel vs. super-pixel-based greenspace mapping along the urban–rural gradient using high spatial resolution Gaofen-2 satellite imagery: A case study of Haidian District, Beijing, China. *Int. J. Remote Sens.* **2017**, *38*, 6386–6406. [CrossRef]

12. Puissant, A.; Rougier, S.; Stumpf, A. Object-oriented mapping of urban trees using Random Forest classifiers. *Int. J. Appl. Earth Obs. Geoinf.* **2014**, *26*, 235–245. [CrossRef]

13. Dey, V.; Zhang, Y.; Zhong, M. A review on image segmentation techniques with remote sensing perspective. In Proceedings of the ISPRS TC VII Symposium—100 Years ISPRS, Vienna, Austria, 5–7 July 2010; Volume XXXVIII. Part 7A.

14. Mathieu, R.; Freeman, C.; Aryal, J. Mapping private gardens in urban areas using object-oriented techniques and very high-resolution satellite imagery. *Landsc. Urban Plan.* **2007**, *81*, 179–192. [CrossRef]

15. Moskal, L.M.; Styers, D.M.; Halabisky, M. Monitoring Urban Tree Cover Using Object-Based Image Analysis and Public Domain Remotely Sensed Data. *Remote Sens.* **2011**, *3*, 2243–2262. [CrossRef]

16. Benz, U.C.; Hofmann, P.; Willhauck, G.; Lingenfelder, I.; Heynen, M. Multi-resolution, object-oriented fuzzy analysis of remote sensing data for GIS-ready information. *ISPRS J. Photogramm. Remote Sens.* **2004**, *58*, 239–258. [CrossRef]

17. Blaschke, T. Object based image analysis for remote sensing. *ISPRS J. Photogramm. Remote Sens.* **2010**, *65*, 2–16. [CrossRef]

18. Kim, M.; Warner, T.A.; Madden, M.; Atkinson, D.S. Multi-scale GEOBIA with very high spatial resolution digital aerial imagery: Scale, texture and image objects. *Int. J. Remote Sens.* **2011**, *32*, 2825–2850. [CrossRef]

19. Carleer, A.P.; Debeir, O.; Wolff, E. Assessment of Very High Spatial Resolution Satellite Image Segmentations. *Photogramm. Eng. Remote Sens.* **2005**, *71*, 1285–1294. [CrossRef]

20. Tian, J.; Chen, D.-M. Optimization in multi-scale segmentation of high-resolution satellite images for artificial feature recognition. *Int. J. Remote Sens.* **2007**, *28*, 4625–4644. [CrossRef]

21. Liu, Y.; Bian, L.; Meng, Y.; Wang, H.; Zhang, S.; Yang, Y.; Shao, X.; Wang, B. Discrepancy measures for selecting optimal combination of parameter values in object-based image analysis. *ISPRS J. Photogramm. Remote Sens.* **2012**, *68*, 144–156. [CrossRef]

22. Witharana, C.; Civco, D.L.; Meyer, T.H. Evaluation of data fusion and image segmentation in earth observation based rapid mapping workflows. *ISPRS J. Photogramm. Remote Sens.* **2014**, *87*, 1–18. [CrossRef]

23. Zhang, X.; Feng, X.; Xiao, P.; He, G.; Zhu, L. Segmentation quality evaluation using region-based precision and recall measures for remote sensing images. *ISPRS J. Photogramm. Remote Sens.* **2015**, *102*, 73–84. [CrossRef]

24. Witharana, C.; Civco, D.L. Optimizing multi-resolution segmentation scale using empirical methods: Exploring the sensitivity of the supervised discrepancy measure Euclidean distance 2 (ED2). *ISPRS J. Photogramm. Remote Sens.* **2014**, *87*, 108–121. [CrossRef]

25. Cardoso, J.S.; Corte-Real, L. Toward a generic evaluation of image segmentation. *IEEE Trans. Image Process.* **2005**, *14*, 1773–1782. [CrossRef] [PubMed]

26. Espindola, G.M.; Camara, G.; Reis, I.A.; Bins, L.S.; Monteiro, A.M. Parameter selection for region-growing image segmentation algorithms using spatial autocorrelation. *Int. J. Remote Sens.* **2006**, *27*, 3035–3040. [CrossRef]

27. Drăguţ, L.; Tiede, D.; Levick, S.R. ESP: A tool to estimate scale parameter for multiresolution image segmentation of remotely sensed data. *Int. J. Geogr. Inf. Sci.* **2010**, *24*, 859–871. [CrossRef]

28. Drăguţ, L.; Csillik, O.; Eisank, C.; Tiede, D. Automated parameterisation for multi-scale image segmentation on multiple layers. *ISPRS J. Photogramm. Remote Sens.* **2014**, *88*, 119–127. [CrossRef] [PubMed]

29. Yang, J.; Li, P.; He, Y. A multi-band approach to unsupervised scale parameter selection for multi-scale image segmentation. *ISPRS J. Photogramm. Remote Sens.* **2014**, *94*, 13–24. [CrossRef]

30. Zhang, X.; Xiao, P.; Feng, X. An Unsupervised Evaluation Method for Remotely Sensed Imagery Segmentation. *IEEE Geosci. Remote Sens. Lett.* **2012**, *9*, 156–160. [CrossRef]

31. Ming, D.; Li, J.; Wang, J.; Zhang, M. Scale parameter selection by spatial statistics for GeOBIA: Using mean-shift based multi-scale segmentation as an example. *ISPRS J. Photogramm. Remote Sens.* **2015**, *106*, 28–41. [CrossRef]

32. Stein, A.; Beurs, K. De Complexity metrics to quantify semantic accuracy in segmented Landsat images. *Int. J. Remote Sens.* **2005**, *26*, 2937–2951. [CrossRef]

33. Karl, J.W.; Maurer, B.A. Spatial dependence of predictions from image segmentation: A variogram-based method to determine appropriate scales for producing land-management information. *Ecol. Inform.* **2010**, *5*, 194–202. [CrossRef]

34. Martha, T.R.; Kerle, N.; van Westen, C.J.; Jetten, V.; Kumar, K.V. Segment Optimization and Data-Driven Thresholding for Knowledge-Based Landslide Detection by Object-Based Image Analysis. *IEEE Trans. Geosci. Remote Sens.* **2011**, *49*, 4928–4943. [CrossRef]

35. Johnson, B.; Xie, Z. Unsupervised image segmentation evaluation and refinement using a multi-scale approach. *ISPRS J. Photogramm. Remote Sens.* **2011**, *66*, 473–483. [CrossRef]

36. Yi, L.; Zhang, G.; Wu, Z. A Scale-Synthesis Method for High Spatial Resolution Remote Sensing Image Segmentation. *IEEE Trans. Geosci. Remote Sens.* **2012**, *50*, 4062–4070. [CrossRef]

37. Zhang, X.; Xiao, P.; Feng, X.; Feng, L.; Ye, N. Toward Evaluating Multiscale Segmentations of High Spatial Resolution Remote Sensing Images. *IEEE Trans. Geosci. Remote Sens.* **2015**, *53*, 3694–3706. [CrossRef]

38. Akçay, H.G.; Aksoy, S. Automatic Detection of Geospatial Objects Using Multiple Hierarchical Segmentations. *IEEE Trans. Geosci. Remote Sens.* **2008**, *46*, 2097–2111. [CrossRef]

39. Esch, T.; Thiel, M.; Bock, M.; Roth, A.; Dech, S. Improvement of Image Segmentation Accuracy Based on Multiscale Optimization Procedure. *IEEE Geosci. Remote Sens. Lett.* **2008**, *5*, 463–467. [CrossRef]

40. Myint, S.W.; Gober, P.; Brazel, A.; Grossman-Clarke, S.; Weng, Q. Per-pixel vs. object-based classification of urban land cover extraction using high spatial resolution imagery. *Remote Sens. Environ.* **2011**, *115*, 1145–1161. [CrossRef]

41. Anders, N.S.; Seijmonsbergen, A.C.; Bouten, W. Segmentation optimization and stratified object-based analysis for semi-automated geomorphological mapping. *Remote Sens. Environ.* **2011**, *115*, 2976–2985. [CrossRef]

42. Kabisch, N.; Haase, D. Green spaces of European cities revisited for 1990–2006. *Landsc. Urban Plan.* **2013**, *110*, 113–122. [CrossRef]

43. Yang, J.; Huang, C.; Zhang, Z.; Wang, L. The temporal trend of urban green coverage in major Chinese cities between 1990 and 2010. *Urban For. Urban Green.* **2014**, *13*, 19–27. [CrossRef]

44. Senanayake, I.P.; Welivitiya, W.D.D.P.; Nadeeka, P.M. Urban green spaces analysis for development planning in Colombo, Sri Lanka, utilizing THEOS satellite imagery—A remote sensing and GIS approach. *Urban For. Urban Green.* **2013**, *12*, 307–314. [CrossRef]

45. Wolch, J.R.; Byrne, J.; Newell, J.P. Urban green space, public health, and environmental justice: The challenge of making cities 'just green enough'. *Landsc. Urban Plan.* **2014**, *125*, 234–244. [CrossRef]

46. Zhang, X.; Xiao, P.; Feng, X. Toward combining thematic information with hierarchical multiscale segmentations using tree Markov random field model. *ISPRS J. Photogramm. Remote Sens.* **2017**, *131*, 134–146. [CrossRef]

47. Baatz, M.; Schäpe, A. Multiresolution Segmentation: An optimization approach for high quality multi-scale image segmentation. In *Angewandte Geographische Informations-Verarbeitung XII*; Strobl, J., Ed.; Wichmann-Verlag: Heidelberg, Germany, 2000; pp. 12–23.

48. Comaniciu, D.; Meer, P. Mean shift: A robust approach toward feature space analysis. *IEEE Trans. Pattern Anal. Mach. Intell.* **2002**, *24*, 603–619. [CrossRef]

49. Arbeláez, P.; Maire, M.; Fowlkes, C.; Malik, J. Contour Detection and Hierarchical Image Segmentation. *IEEE Trans. Pattern Anal. Mach. Intell.* **2011**, *33*, 898–916. [CrossRef] [PubMed]

50. Drăguţ, L.; Eisank, C. Automated object-based classification of topography from SRTM data. *Geomorphology* **2012**, *141–142*, 21–33. [CrossRef] [PubMed]

51. Tucker, C.J. Red and photographic infrared linear combinations for monitoring vegetation. *Remote Sens. Environ.* **1979**, *8*, 127–150. [CrossRef]

52. Zhang, Y.J. A survey on evaluation methods for image segmentation. *Pattern Recognit.* **1996**, *29*, 1335–1346. [CrossRef]

53. Van Rijsbergen, C.J. *Information Retrieval*, 2nd ed.; Butterworth-Heinemann: Newton, MA, USA, 1979; ISBN 978-0-408-70929-3.

Article

Improving Ecotope Segmentation by Combining Topographic and Spectral Data

Julien Radoux [1,*,†], Axel Bourdouxhe [2,†], William Coos [2,†], Marc Dufrêne [2,†] and Pierre Defourny [1,†]

1 Earth and Life Institute, Université catholique de Louvain, 1348 Louvain-la-Neuve, Belgium; pierre.defourny@uclouvain.be
2 Biodiversity and Landscape Unit, Gembloux Agro-Bio Tech, Université de Liège, 5030 Gembloux, Belgium; axel.bourdouxhe@uliege.be (A.B.); william.coos@alumni.ulg.ac.be (W.C.); marc.dufrene@uliege.be (M.D.)
* Correspondence: julien.radoux@uclouvain.be; Tel.: +32-(0)10-479257
† These authors contributed equally to this work.

Received: 14 January 2019; Accepted: 30 January 2019; Published: 11 February 2019

check for
updates

Abstract: Ecotopes are the smallest ecologically distinct landscape features in a landscape mapping and classification system. Mapping ecotopes therefore enables the measurement of ecological patterns, process and change. In this study, a multi-source GEOBIA workflow is used to improve the automated delineation and descriptions of ecotopes. Aerial photographs and LIDAR data provide input for landscape segmentation based on spectral signature, height structure and topography. Each segment is then characterized based on the proportion of land cover features identified at 2 m pixel-based classification. The results show that the use of hillshade bands simultaneously with spectral bands increases the consistency of the ecotope delineation. These results are promising to further describe biotopes of high ecological conservation value, as suggested by a successful test on ravine forest biotope.

Keywords: GEOBIA; biodiversity; LIDAR; orthophoto; segmentation; classification; biotope distribution model

1. Introduction

1.1. Context

In order to mitigate biodiversity loss and destruction of ecosystems with heritage value around the world, we have to know where biodiversity hotspots and threatened areas are located. Facing the actual threats and due to a big extinction rate, the urgency leads to a race to become aware and map theses area before they don't exist anymore. This logic was followed at many scales. Worldwide, biodiversity hotspots were identified and outlined in order to prioritize conservation actions [1,2].

At the European scale, two directives have defined the need for the conservation of habitats and species with the adoption of appropriate measures. They allow to give a protection status for species and biotopes of interest, but also defining protected areas corresponding to species habitats or group of biotopes. Within this Pan-European ecological network known as "Natura 2000 network" of special areas of conservation, natural habitats will be monitored to ensure the maintenance or restoration of their composition, structure and extent [3].

Monitoring the evolution of the territory (land cover, habitats of species, biotopes, ...) is an essential activity to identify major changes, economic, social and environmental issues but also to assess the impact of public policies and private initiatives. This monitoring is held by each country and requires a large amount of data, mainly obtained through field surveys having a high financial and time cost. These mapping results are used to mitigate problems such as conservation measures of the kind at national and local level [4], planning and development of green infrastructure [5], agro-environmental assessments [6], landscape changes monitoring [7,8], ecological forest management [9] or identification of ecosystem services [10,11]. However, existing maps are often limited to categorical land cover characterization which does not provide a precise legend for habitat and biotope types and are hardly interoperable. Innovative remote sensing products could, however, facilitate the status monitoring and the detailed characterization of large areas, even sometimes for fine scale quality indicators [12]. While it does not replace field data collection, remote sensing integration could thus be a first step towards a more cost effective monitoring of natural habitats [13].

Because of the limitations of remote sensing, habitat suitability mapping and biotope prediction models are necessary to fill the gaps of field observation for biodiversity monitoring. Nevertheless, remotely sensed data are of paramount importance in providing some spatially comprehensive information that is necessary to the prediction over large regions [14,15]. In this context, models are often based on regular grids linked with permanent structured inventories. However, with the democratization of geopositioning devices and the rise of citizen science, the precision of the observation has tremendously increased. An alternative approach to grid based habitat and biotope prediction could therefore emerge with a landscape partitioning into ecologically meaningful irregular polygons.

1.2. Remote Sensing for Ecotope Mapping

Previous studies showed that irregular polygons were supportive of habitats model that outperformed the standard grid-based approach with more than half of the investigated species [16]. This partition of the landscape into spatially consistent regions can be related to the concepts of ecotopes [17] or of land use management units [18]. Ecotopes are the smallest ecologically distinct landscape features in a landscape mapping and classification system. Mapping ecotopes therefore enables the measurement of ecological patterns, process and change [19] with much more details than categorical land cover classes or continuous field of a single class land cover feature.

Ecotope maps are often created by overlaying a large number of components, such as physiotope (topographic and soil features) and biotope (vegetation) layers [20,21]. As a result, ecotope maps are classified into hundreds of types and dozens of groups by combining biological and geophysical variables [22]. Furthermore, the different scales and precision of the boundaries of the overlaid thematic layers may create many artifacts which need to be handled with advanced conflation rules.

Alternatively, Geographic Object-Based Image Analysis can be used to delineate spatial regions by grouping adjacent pixels into homogeneous areas according to the objectives of the study [23,24]. For biodiversity research, image segmentation has been used to automatically derive homogeneous vegetation units based on spectral [25] or a combination of spectral and structural (height) information [16,26]. These approaches helped to reduce the number of polygons and improved the matching of those polygons with entities derived from the field.

On the other hand, GEOBIA was also used to delineate physiotopes, which can then be overlaid with land cover polygons to derive meaningful spatial regions [18] or directly used to map aquatic habitats [27]. The delineation of physiotopes is a difficult task to assess because their definition depends on the purpose of the study [28]. Different GEOBIA methods have therefore been developed, based on curvature indices [18,29], decision rules using elevation and slope [30], network properties [28] or a large set (70) of indices including slope, aspect and various texture indices [27]. However, the methods designed for terrestrial landscapes focused on global to regional scales, where the relative position (ridge, side or valley) plays a major role for the classification and do not directly take the orientation of the slope into account.

Our study aims at improving the large scale delineation of ecotopes applied on ecological modeling in Delangre et al. [16]. Our hypothesis is that this improvement can be achieved by simultaneously processing the topographic information from a LIDAR DEM and the vegetation structure information from optical image and LIDAR DHM. Topography is indeed a major driver of other abiotic components such as soil properties (which is more difficult to obtain at high precision) [31,32] or insulation (depending on the orientation of the slope) [33].

2. Data and Study Area

The study area is located in the Walloon region (Southern part of Belgium). This is a very fragmented landscape including coniferous forests (mainly spruce and other sempervirent species), deciduous broadleaved forests (mainly oaks and beeches), crop fields, natural and managed grasslands, peatlands, small water bodies, extraction areas as well as dense and sparse urban fabrics.

There are no mountainous terrains in Belgium, but a topography that is mainly driven by a dense hydrological network. In order to test our hypothesis, the experimental study focuses on the ravine maple stands, which grow on relatively steep slope and rocky soil. This biotope is particularly sparse, but at least present in five of the biogeographical regions of Wallonia. A rectangular study area (Figure 1) was delineated to include the majority of these biotopes present in Belgium. This region is relatively flat (slopes smaller than 7 percents) except in the valleys.

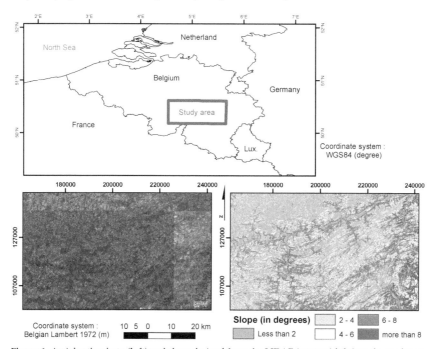

Figure 1. Aerial orthophoto (**left**) and slope derived from the LIDAR images (**right**) on the study area.

Two types of input data were available in the study area. First, a mosaic of ortho-rectified aerial photographs upscaled to 2 m resolution and including four spectral bands; Second, a LIDAR point cloud dataset rasterized at 2 m resolution.

The aerial photographs cover the entire study area. This coverage was done with several flights between March and April 2015. Image acquisition included four spectral bands (blue, green, red and near-infrared) at a spatial resolution of 0.25 m. The images available for the analysis were already ortho-rectified, mosaicked and rescaled in bytes. In order to avoid too much local heterogeneity, which would affect classification process, the original images were resampled at 2 m resolution using the mean values of all contributing pixels.

The LIDAR dataset was acquired in spring 2013 and 2014. The minimum sampling density is of 0.8 points per square meter. First and last returns were used to extract the ground elevation and the vegetation canopy height. In addition, this dataset required specific mathematical morphology analysis in order to remove some artifacts: a gray scale opening was applied in order to remove power lines. A digital elevation model (DEM) and a Digital Surface Model (DSM) of the vegetation were derived from the last and first returns, respectively. A Digital Height Model (DHM) was then obtained by subtracting the DEM from the DSM.

In addition to the remote sensing data, a vector database describing the biotopes inside Belgian protected areas from the European network of natural sites (NATURA2000) was available. This database was produced by the Walloon administration for Nature and Forest based on expert knowledge and exhaustive field inventories (all polygons). It is considered as the best available information about ravine forests in Wallonia and was therefore used as a reference. In order to ensure and improve the reliability of this map, polygons with visible clear cuts on the 2015 othophotos were manually removed from this reference.

3. Method

The core of the proposed process is the simultaneous segmentation of the topographic, spectral and height information. The resulting image segments are then enriched by computing a set of attributes based on remote sensing and ancillary data. The potential of the proposed method to automatically delineate meaningful spatial regions is assessed based on two expected properties of the ecotopes: a large homogeneity and the ability to build high performance ecological models. These steps are summarized on Figure 2.

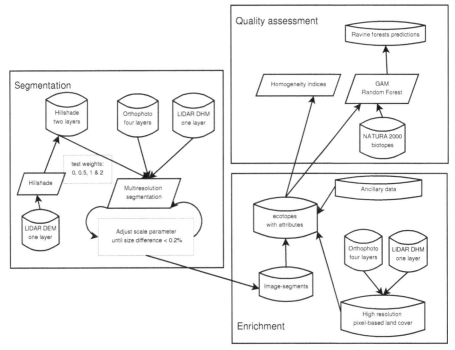

Figure 2. Overall flowchart of the proposed method.

3.1. Automated Ecotope Delineation

The three variables of interest to discriminate ecological function at the scale of the analysis are the land cover, the topography and the soil type. However, the available soil type information was not precise enough and could be partly inferred by the topography. We therefore focused on variables that could be directly inferred by remote sensing: topography and land cover.

The multiresolution segmentation algorithm [34] was used to automatically delineate ecotopes. This algorithm can be tuned by a set of four parameters: the scale, the weight of the raster layer, the shape and the compactness. The scale parameter defines the maximum acceptable value of the change of heterogeneity when merging two neighboring image-segments. Increasing the scale parameter therefore increases the size of the image-segments. The weight of the layers defines how much each raster layer will contribute to the heterogeneity difference of the merged image-segments as shown in Equation (1):

$$h_{diff} = \sum_L w_L (n_1 (h_{mergedL} - h_{1L}) + n_2 (h_{mergedL} - h_{2L}) / \sum_L w_L \qquad (1)$$

where h_{diff} is the total heterogeneity difference after merging based on the raster layers, w_L is the weight of each raster layer, $h_{mergedL}$ is the heterogeneity of image-segments 1 and 2 for layer L; n_1 and n_2 are the number of pixels in image-segments 1 and 2; h_{2L} and h_{2L} are the heterogeneity indices of image-segments 1 and 2. Then, the shape parameter defines the proportion of the heterogeneity index that is based on the shape of the image-segment. Increasing the shape parameter therefore reduces the contribution of a large heterogeneity difference after merging the image-segment. The compactness parameter determines if this shape index should favor compact image-segments (similar to a disk) or smooth image-segments (similar to a rectangle).

The efficiency of a segmentation combining LIDAR height and multispectral image had already been proven [26]. Our working hypothesis is that simultaneously combining the topographic information with the spectral values of the orthophotos and the DHM derived from the LIDAR would improve the delineation of the ecotopes. The segmentation results are therefore compared with different weights to the topographic information with respect to the other layers. For the sake of a fair comparison, the average size of the image-segments is fixed to approximately 2 ha (Two hectare on average corresponds to smallest ecological management units according to a group of users including biodiversity researchers and managers.) To do so, the composite image was first segmented with a scale parameter of 50, a shape parameter of 20% and a compactness of 100%. The shape parameter was then reduced to 10% and a larger scale parameter was obtained using binary search algorithm with a tolerance of 0.2% on the total number of polygons obtained on the reference image segmentation (that is 318,380). Apart from the size that was fixed after the first segmentation, no other optimization of the segmentation was performed. The only difference between the segmentation is therefore the weight of the topographic component that is being tested with values of zero (only spectral and structural information), 0.5, 1 and 2 (increasing the influence of topographic information).

Including topography in segmentation required a transformation of the DEM data to highlight the different slope types and identify breaks. Because the segmentation algorithm is based on the minimization of the variance inside each image-segment, using DEM values would indeed tend to create many linear spatial regions along contour lines in areas of steep slopes, even if the slope is constant. Previous studies used the slope together with some curvature indices [18]. This is interesting for pedomorphic mapping, but (i) it then relies on arbitrary window size to compute minimum and maximum curvature and (ii) both sides of ridge and valley lines are in the same segment despites different sun illumination. In the case of ecotopes, the slope and the aspect of the slope are therefore more closely related to the functionnal homogeneity However, slope aspect could not be used by the segmentation algorithm because (i) it is undefined when the slope is null and (ii) it is a circular metric that jumps from 360 degrees to 0 degree for the same azimuthal direction. For those reasons, Janowski et al. [27] used easting and northing instead of azimuth. For the ecotopes, two synthetic hillshade maps were derived along the North-South and the East-West transects using Equation (2), because this is the variable that is the most directly linked with the potential solar energy.

$$hillshade = 255 \times ((cos(SZA) \times cos(Slope)) + (sin(SZA) \times sin(Slope) \times cos(SAA - Aspect))) \quad (2)$$

where SZA and SAA are the hypothetical sun zenithal and azimuthal angles, respectively, and *Slope* and *Aspect* are derived from the DEM using a 3-by-3 moving window. The use of 3-by-3 windows corresponds to local hillshade at a high spatial resolution (2 m), so that only pixels with similar hillshade values are likely grouped together by the segmentation algorithm. The shape parameter of 20% that is used in the segmentation process aims at preserving the compacity of the image segment when isolated pixels have a markedly different orientations than their surroundings, but image-segments are expected not to merge when there is a change of slope.

In practice, synthetic hillshade maps were created by setting a large sun zenith angle (75°) for four sun azimuth angles (0°, 90°, 180°, 270°). The difference between the results of both pairs of opposite theoretical sun azimuth angles were then computed. Cast shadows were ignored in this process because the aim of the hillshade is only to provide a continuous topographic characterization. As can be seen on Figure 3, the values of the hillshade are equal on flat surfaces and on slopes oriented with 45° or 135° azimuths. In the case of flat areas, the value in two opposite directions is indeed equal, so that their difference is zero for all azimuths. In the other case, the values are either positive or negative and they are equal for the orthogonal direction because 45° is the bisector of those azimuthal angles.

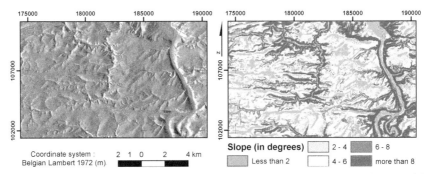

Figure 3. False color composite of the hillshades along the North-South and East-West directions (**left**) and slope derived from the LIDAR images (**right**) on a subset of the study area. Shades of grey indicate that the hillshade values in the two orthogonal directions are equal while colored areas highlight the differences between the two directions.

3.2. Quality Assessment

For the sake of ecological models, image-segments are enriched based on the proportion of land cover features that they contain as well as various soil and contextual attributes [16]. With all attributes being derived from external databases, the quality assessment focuses on the homogeneity of the image-segments (which is a key feature for ecotopes) (Section 3.2.2) and the ability to run performant ecological models (Section 3.2.3).

Due to the lack of other up-to-date high resolution land cover map of the Walloon region at the time of the study, a high resolution pixel-based land cover map was produced in order to characterize the ecotopes and build some of their homogeneity indices. While the production of this high resolution land cover database is out of the scope of this paper, it is briefly described in Section 3.2.1.

3.2.1. High Resolution Pixel-Based Land Cover

A Bayesian classifier with automated training sample extraction method [35] was used to classify 8 land cover types: bare soil, artificial, grassland, crops, coniferous, broadleaved, water and shrubs.

The input image was based on the same datasets as the segmentation: the 4 spectral bands of the aerial photograph and the height information extracted from LIDAR. The *a priori* probability was computed based on the frequency of each land cover type within two height classes (below and above 50 cm). Because of the high reliability of the LIDAR DHM, this step was particularly useful to discriminate forests, shrubs and buildings from the other land covers. The training dataset was compiled based on existing datasets covering the study area, including a 2007 land cover map from the Walloon Region, Open Street Map data [36] and forest inventory data from the Nature and Forest Department. The results were then consolidated with a crop mask in order to discriminate grassland and cropland. Furthermore, the classification of forest types was consolidated in the homogeneous region thanks to a classification of Sentinel-2 cloud-free images of early spring and mid summer 2016 (assuming that the forest type does not change from year to year and excluding clear cuts from the analysis).

For the validation, a simple random sample of 700 points was photointerpreted on the orthophoto with a geolocation tolerance of 5 m and ambiguous points were verified on the ground. The estimated overall accuracy of the consolidated product (93% with a 95% confidence) was above the other products, therefore it was considered as our best reference in the frame of this paper.

3.2.2. Homogeneity Measures

In order to test the hypothesis of this study, different homogeneity indices have been computed. Those indices look at the homogeneity from land cover (based on the high resolution land cover layer), from the topography and from the soil types. They are compared with an arbitrary regular grid with the same cell area than the average polygon size, which provides a reference considering the segmentation ratio. Because of the specific interest towards a biotope that is mainly present in areas of steep slopes, the homogeneity indices were not only computed for all the study area, but also for a subset composed of the polygons with an average slope above 10 degrees.

Giving more weight to the topographic bands could affect the homogeneity in terms of land cover delineation. In order to control a potential loss of land cover homogeneity, the average purity level was computed for each segmentation. The proportion of each land cover class was computed inside each polygon based on the high-resolution pixel-based land cover classification presented in Section 3.2.1. The purity index is then defined as the average of the maximum values of land cover proportions of each image-segment.

From the topographic point of view, the primary variable of interest is the slope. The slope was measured on a smoothed version of the 2 m DEM in order to remove micro-topography effects and to remove artifacts due to the noise of the dataset. Because the slope is a quantitative variable, its heterogeneity was estimated using the standard deviation (STD) inside each polygon. For the aspect of the slope, standard deviation could not be used because of the break between 0 and 360°. The azimuth values were therefore converted into nine categories, including in eight directions (North, North-East, East, South-East, South, South-West, West and North-West) plus one class for the flat areas (where the aspect is undefined). The purity index of these nine categories is then computed like in the case of the land cover.

Finally, an independent data source was also considered: the soil map. The purity index for soil drainage classes and soil depth classes was used as an additional indirect indicator of the polygon homogeneity. Those two soil classes were derived from the digital soil map of the Walloon region. The precision of this map corresponds to a scale of 1/25,000, which is coarser than the polygons delineation, but this uncertainty affects all polygon boundaries in a similar way.

3.2.3. Biotope Models

In addition to the homogeneity measures, a fitness to purpose analysis was implemented. The sensitivity of two state-of-the-art algorithms, namely Random Forest (RF) and Generalized Additive Model (GAM), has been tested for the detection of ravine maple forests. Each model was calibrated using the same workflow for each of the segmentation results.

First, a large set of attributes have been derived from existing database and GIS analysis. This set includes bioclimatic variables interpolated from Worldclim [37], soil variables, topographic variables and land cover variables obtained by zonal statistics within each ecotope. Those variables have been selected based on expert knowledge and their contribution to habitat suitability models have been assessed in a previous study [16].

Calibration and validation polygons were then selected by crossing the ecotope database with the polygons of the NATURA 2000 database. An ecotope was labeled as a ravine maple forest biotope if more than half of its area was covered by its equivalent in the Natura 2000 cartography. To obtain a presence/absence dataset, ecotopes matching with ravine maple forests were considered as presence, while ecotopes matching with any other forest biotope were considered as an absence.

Different quality indices were used to validate the model, including the Overall Accuracy (OA) and the Area Under the Curve (AUC) of the model as well as producer and user accuracy (PA and UA) of the optimal binary classification between ravine forest and other forest biotope. In order to evaluate the accuracy of the model to detect ravine forests among all other biotopes, another overall accuracy was calculated taking into account all surfaces covered by Natura 2000 surveys (OA_Tot). Those indices were computed for the validation polygons which have been separated from the rest of the dataset before the calibration step. In order to provide an unbiased estimate of the correctly classified areas, the ecotope polygons were used as sampling units and their areas were taken into account [38]. The optimal binary classification was automatically determined based on the best compromise between sensitivity and specificity.

4. Results

By design, approximately 318400 image-segments were automatically created in the study area (with a range of 500 polygons, that is less than 0.2 percent). A visual check did not catch any macroscopic errors, but revealed most of the topographic features hidden by the vegetation on the aerial image. Figure 4 shows a subset of the segmentation result, highlighting the impact of the topography on the image-segments created inside patches of homogeneous land cover. As expected, areas of homogeneous slope are well delineated in addition to the land cover induced partitioning. Furthermore, the limits of the ecotopes are consistent with the pattern of slope curvature, which were not used for the segmentation.

Quantitative results related to the homogeneity of the image-segments are summarized in Tables 1 and 2. Overall, the advantage of automatically partitioned landscape against a regular grid of the same size is obvious. The results indeed show that the heterogeneity of the topographic attributes decreases and the separability of ecotopes increases when the partition of the landscape is determined by topography and land cover. As shown in Figure 2, the results within the subset of polygons with a slope above 15 percent further highlight the differences where the terrain plays a bigger role in the definition of the polygons.

Table 1. Homogeneity of the image-segment as a function of the segmentation weights. The grid is composed of squares with the same area as the average of image-segments. Large purity values and low average variance of the slope indicate a good segmentation.

	0 (No Topographic Layers)	0.5	1	2	Grid
Slope variance	4.21	4.00	3.90	3.83	4.82
Aspect purity	94.4	94.4	94.5	94.5	94.3
Soil depth purity	82.8	82.8	84.0	83.1	79.9
Soil drainage purity	80.1	80.9	81.3	81.7	80.4
Land cover purity	75.9	76.5	76.6	76.4	72.2

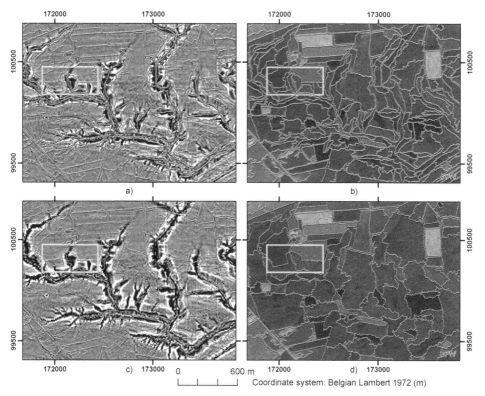

Figure 4. Image-segment boundaries (in red) overlaid on the curvature of the DEM at 10 m resolution (**left**) and the orthophoto from the Walloon Region (**right**, copyright SPW 2015). The images at the top (**a,b**) display the segmentation with the topographic bands (weight of 1); the images at the bottom (**c,d**) display the segmentation without the topographic bands. The green rectangle highlights an area where the land use boundaries follow the topography.

Table 2. Homogeneity of the subset of image-segments with a slope greater than 15 percent. The grid is composed of squares with the same area as the average of image-segments. Large purity values and low average of slope variance indicate a good segmentation.

	0 (No Topographic Layers)	0.5	1	2	Grid
Slope variance	10.6	8.1	7.1	6.2	11.6
Aspect purity	93.7	95.2	95.9	96.4	93.7
Soil depth purity	80.2	81.9	82.7	83.5	75.6
Soil drainage purity	79.7	81.8	82.4	82.5	78.3
Land cover purity	69.4	75.0	75.4	76.4	64.8
Mean area (m^2)	20,466	17,379	16,432	15,577	19,016

The analysis of the predictive model indicate that the ecotopes are appropriate mapping units to map ravine forest in the study area. The overall accuracy of the best model is indeed 99.9% (Table 3). However, this value does not completely reflect the errors of the model because ravine forests are rare in the study area and specific to the polygons with a majority of broadleaved trees. Additional indices measured on a subset of the ecotopes with a majority of broadleaved trees are therefore more relevant to compare the different scenarios. On this subset, the use of topographic information to delineate polygons had a significantly positive impact on the results of the models. This confirms the results obtained by homogeneity measures (Tables 1 and 2). However, the best prediction is achieved for the

GAM model when the weight of the topographic information is equal to the weight of the spectral information and the performances of the model decrease with the weight of 2.

Table 3. Results of the different models of ravine forest predictions with respect to the weight of topography in the segmentation. Matches correspond to the number of ecotopes matching at more than 50 percent with biotopes survey polygons. The next rows show the different quality indices including the overall accuracy of ravine forest mapping in the study area (OA_{Tot}), the Overall Accuracy (OA), Area Under the Curve (AUC), Producer Accuracy (PA) and User Accuracy (UA) of the ravine forests for each model based on ecotopes covered by a majority of broadleaved trees.

	0	0.5	1	2
Matches	17	60	87	109
RF OA_{Tot}	99.7	99.8	99.8	99.7
RF OA	93.2	94.7	95.5	92.7
RF AUC	79.6	97.1	96.8	94.3
RF PA	77.9	97.0	95.3	92.7
RF UA	8.90	18.9	25.3	16.1
GAM OA_{Tot}	99.8	99.8	99.9	99.8
GAM OA	96.1	95.7	97.3	95.2
GAM AUC	81.4	95.6	97.6	95.2
GAM PA	77.2	93.1	97.0	92.7
GAM UA	15.0	21.9	37.2	22.5

The number of matching polygons increases when the segmentation uses more topographic components (Table 3). This is due to the fact that the polygon boundaries match the Natura 2000 boundaries closer than in the case without topographic contributions, but also because the polygons become on average smaller in rugged terrain when the topography is taken into account, as shown in Table 2.

5. Discussion

This study demonstrates that automated image segmentation simultaneously combining topographic information from LIDAR with the spectral information from optical sensors provides ecologically relevant polygons. Two facets of the results are discussed in this section: the technical quality of the results and the usefulness of the model for biodiversity studies.

5.1. Consistency of the Polygons

The objective of this paper was to build homogeneous polygons that would better match the concept of ecotopes than a delineation solely accounting for the land cover. While it was foreseen that the addition of topographic information to the segmentation process would reduce the topographic heterogeneity, the increased land cover homogeneity was surprising. This could be due to the long term land management practices that optimized land use based on the topography (for example, most crop fields are located on flat areas while steep slope are mainly covered by forests). Such patterns have been observed in the landscape, but the causality should be further investigated.

The results of the model shows that including topographic information improves the correspondence between ecotopes polygons and the field mapping of ravine forest biotope. The increased number of matches is partly due to the reduction of the average polygon size inside rugged terrain (about 25% for the weight of 2). The reduction of the size is however not sufficient to explain the large increase in the number of matches. This could be better explained by the fact that presence of biotope is highly dependent on the topographic situation. Indeed, contrasted situations such as south or north hillside leads to very different abiotic features. Furthermore, even a small difference of slope leads to different water intakes leading also to different vegetation communities.

Even if the model weren't created in the best conditions due to the scarcity of the biotope, we can stress that we see a large leap in the AUC of the models (more than 16% for both RF and GAM) by adding topographic data in the segmentation process. Concerning the model accuracy, the big increase observed by adding topographic data is consistent with the improvement observed by the heterogeneity indices. However, we can see that the indices of the models don't follow the same trend than the heterogeneity indices and are less correlated with the level of contribution of topographic data. This is probably due to the fact that we model a rare biotope with scarce presence data. Thus, the overall quality of the models is sensitive to the polygons selected in the calibration/validation process.

In this study, the average size was selected according to user requirements and fixed in order to fairly compare the contributions of the models. However, the size could also be optimized based on data driven features [39,40] In this case, the weight of the topography, the vegetation height and the spectral values of the images should be considered to determine an optimal size. However, the observed difference between the trend in terms of ecotope purity and the trend of the fitness-to-purpose analysis should be further investigated as a potential issue to the use of a single optimization criteria.

On the other hand, landscape and landform analysis very much depends on the scale of the process being addressed and a hierarchical approach could help to extract additional characteristics from the landscape. For instance, elevation and curvature are important features at coarser scale to identify ridges or valleys. However, ridges and valley landforms include two sides facing opposite directions, which is not homogeneous in terms of insulation. The use of hillshade at local scale obviously placed the emphasis on potential insulation, but it also split image-segments in places of strong positive or negative curvature, which contributed to the improvement of characteristic soil properties. Identifying pattern from another scales could however be necessary to cover the processes that occur in more mountainous areas.

5.2. Usefulness of Biotope Models

The high (above 95% for both GAM and RF) AUC of the best models shows that models based on ecotope polygons are very consistent. However, despite the high specificity and sensitivity of the models, the user accuracy of the ravine forest class is very low. This low user accuracy can be explained by three different factors.

First, the ravine forests are rare in the study area. As a result, even a very small proportion of false positive had a strong impact on the user accuracy. Nevertheless, the absolute number of forest remains small, so that the models could help field prospecting by narrowing the search area to fewer than five percents of the total surface of forests.

Second, the models were used to test a concept and compare different segmentation strategies, but they could be tuned for other specific purposes. The selected optimization criteria, based on the sensitivity and the specificity, favored a solution that maximized the sensitivity because the specificity computed on a large proportion of absences was rapidly very high (above 97%), then slowly increased. This type of optimization is particularly useful for restraining the surveyed area in order to exhaustively map a specific biotope. Another threshold for binary classification could seek an optimum between producer's and user's accuracy based on the F-Score. With this alternative, the UA would be 0.63 and the PA would be 0.65, which is a good compromise for a generic result but less interesting for the identification of unknown biotopes of high biological interest.

Third, the model is limited by the available data, which does not replace field based observations. For instance, typical ferns in the understory are not visible by remote sensing. On the other hand, the ecotope might include all conditions for the development of maple ravine forests, but a different type of forest could have developed because of historical land management of the ecotope. From this point of view, a substantial proportion of the false positives could be considered priority zones for biotope restoration.

Nevertheless, values of user accuracy that we are discussing about are dependent of the correctly classified ravine forest among other broadleaved forests.If we take a look at total overall accuracy values, they show an excellent prediction of ravine forest in our study area therefore moderating the poor user accuracy previously discussed.

Referring to the field data, forest communities don't follow a logic of tangible frontier, but they look like a gradient of vegetation communities mixing with another one. Limits are vague, but the conceptual model of the geographic database uses crisp boundaries to represent those transition areas. This conceptual mapping model into the so-called spatial regions is not specific to the GEOBIA, but it is also performed on the field when a boundary has to be drawn. While about half of the boundaries are matching the boundaries of the reference polygons with precision close to 10 m, diverging delineation occurs on the other half of the boundaries (Example on Figure 5). An independent field campaign was unable to undoubtedly and consistently arbitrate between the two datasets. Despite its limitations, the repeatability of the automated image segmentation makes it very useful in prospective studies or to guide interpretation on the field. However, the use of sharp boundaries could be an issue to represent gradients of vegetation or to be associated with punctual observations. It is therefore of paramount importance to remember that the proposed mapping strategy is a model used to represent the landscape in a way that closely matches the definition of biotopes from the field, but that there is no universal representation of nature.

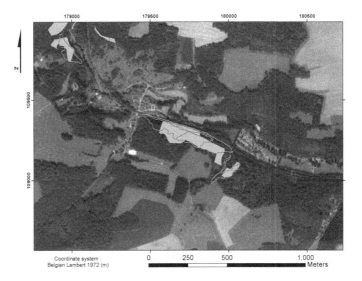

Figure 5. Divergences of delineation of ravine forest in the Natura 2000 database (green polygons) and the automated segmentation (red outlines).

6. Conclusions

This study demonstrates that hillshade layers can be used simultaneously with spectral information to improve the automated delineation of ecotopes in a GEOBIA framework. The AUC of predictive ecological models was improved by 15% when the ecotopes were delineatd using these topographic layers. Furthermore, the inclusion of topographic features in the segmentation process also improved the purity in terms of land cover, probably due to the indirect impact of the topography on the land use in the study area.

The good results at small scale factor suggests that the proposed GEOBIA workflow could be tested at larger scale factor in combination with curvature indices in order to generate homogeneous landforms with minimal arbitrary decisions.

Remote Sens. **2019**, *11*, 354

Author Contributions: Conceptualization, J.R., A.B., W.C., M.D. and P.D.; Formal analysis, J.R. and A.B.; Methodology, J.R., A.B. and W.C.; Supervision, M.D. and P.D.; Validation, J.R. and A.B.; Writing—original draft, J.R., A.B. and W.C.; Writing—review and editing, J.R., A.B., M.D. and P.D.

Funding: This research was funded by Fédération Wallonie-Bruxelles in the frame of the Lifewatch-WB project.

Acknowledgments: The authors thank the three reviewers for their constructive comments.

Conflicts of Interest: The authors declare no conflict of interest.

Dataset: Extended database is available on http://maps.elie.ucl.ac.be/lifewatch/ecotopes.html.

Abbreviations

The following abbreviations are used in this manuscript:

AUC	Area Under the curve
DEM	Digital Elevation Model
DHM	Digital Height Model
GAM	Generalized Additive Model
GEOBIA	Geographic Object-Based Image Analysis
PA	Producer's accuracy
OA	Overall accuracy
RF	Random Forest
STD	Standard Deviation
UA	User's accuracy

References

1. Mittermeier, R.A.; Turner, W.R.; Larsen, F.W.; Brooks, T.M.; Gascon, C. Global Biodiversity Conservation: The Critical Role of Hotspots. In *Biodiversity Hotspots: Distribution and Protection of Conservation Priority Areas*; Zachos, F.E., Habel, J.C., Eds.; Springer: Berlin/Heidelberg, Germany, 2011; pp. 3–22.
2. Myers, N.; Mittermeier, R.A.; Mittermeier, C.G.; da Fonseca, G.A.B.; Kent, J. Biodiversity hotspots for conservation priorities. *Nature* **2000**, *403*, 853–858. [CrossRef] [PubMed]
3. Ostermann, O.P. The need for management of nature conservation sites designated under Natura 2000. *J. Appl. Ecol.* **1998**, *35*, 968–973. [CrossRef]
4. Loidi, J. Preserving biodiversity in the European Union: The Habitats Directive and its application in Spain. *Plant Biosyst.* **1999**, *133*, 99–106. [CrossRef]
5. Wells, M.; Timmer, F.; Carr, A. Understanding Drivers and Setting Targets for Biodiversity in Urban Green Design. In *Green Design: From Theory to Practice*; Yeang, K., Spector, A., Eds.; Black Dog Publishing: London, UK, 2011.
6. Donald, P.F.; Evans, A.D. Habitat connectivity and matrix restoration: The wider implications of agri-environment schemes. *J. Appl. Ecol.* **2006**, *43*, 209–218. [CrossRef]
7. Bryn, A. Recent forest limit changes in south-east Norway: Effects of climate change or regrowth after abandoned utilisation? *Norsk Geogr. Tidsskr.-Nor. J. Geogr.* **2008**, *62*, 251–270. [CrossRef]
8. Bunce, R.; Metzger, M.; Jongman, R.; Brandt, J.; De Blust, G.; Elena-Rossello, R.; Groom, G.B.; Halada, L.; Hofer, G.; Howard, D.; et al. A standardized procedure for surveillance and monitoring European habitats and provision of spatial data. *Landsc. Ecol.* **2008**, *23*, 11–25. [CrossRef]
9. Pokharel, B.; Dech, J.P. An ecological land classification approach to modeling the production of forest biomass. *For. Chron.* **2011**, *87*, 23–32. [CrossRef]
10. Maes, J.; Egoh, B.; Willemen, L.; Liquete, C.; Vihervaara, P.; Schägner, J.P.; Grizzetti, B.; Drakou, E.G.; La Notte, A.; Zulian, G.; et al. Mapping ecosystem services for policy support and decision making in the European Union. *Ecosyst. Serv.* **2012**, *1*, 31–39. [CrossRef]
11. Egoh, B.; Drakou, E.G.; Dunbar, M.B.; Maes, J.; Willemen, L. *Indicators for Mapping Ecosystem Services: A Review*; European Commission, Joint Research Centre (JRC): Ispra, Italy, 2012.
12. Spanhove, T.; Borre, J.V.; Delalieux, S.; Haest, B.; Paelinckx, D. Can remote sensing estimate fine-scale quality indicators of natural habitats? *Ecol. Indic.* **2012**, *18*, 403–412. [CrossRef]

13. Borre, J.V.; Paelinckx, D.; Mücher, C.A.; Kooistra, L.; Haest, B.; De Blust, G.; Schmidt, A.M. Integrating remote sensing in Natura 2000 habitat monitoring: Prospects on the way forward. *J. Nat. Conserv.* **2011**, *19*, 116–125. [CrossRef]

14. Guisan, A.; Zimmermann, N.E. Predictive habitat distribution models in ecology. *Ecol. Model.* **2000**, *135*, 147–186. [CrossRef]

15. Osborne, P.E.; Alonso, J.; Bryant, R. Modelling landscape-scale habitat use using GIS and remote sensing: A case study with great bustards. *J. Appl. Ecol.* **2001**, *38*, 458–471. [CrossRef]

16. Delangre, J.; Radoux, J.; Dufrêne, M. Landscape delineation strategy and size of mapping units impact the performance of habitat suitability models. *Ecol. Inform.* **2017**, *47*, 55–60. [CrossRef]

17. Ellis, E.C.; Wang, H.; Xiao, H.S.; Peng, K.; Liu, X.P.; Li, S.C.; Ouyang, H.; Cheng, X.; Yang, L.Z. Measuring long-term ecological changes in densely populated landscapes using current and historical high resolution imagery. *Remote Sens. Environ.* **2006**, *100*, 457–473. [CrossRef]

18. Gerçek, D. A Conceptual Model for Delineating Land Management Units (LMUs) Using Geographical Object-Based Image Analysis. *ISPRS Int. J. Geo-Inf.* **2017**, *6*, 170. [CrossRef]

19. Chan, J.C.W.; Paelinckx, D. Evaluation of Random Forest and Adaboost tree-based ensemble classification and spectral band selection for ecotope mapping using airborne hyperspectral imagery. *Remote Sens. Environ.* **2008**, *112*, 2999–3011. [CrossRef]

20. Haber, W. Basic concepts of landscape ecology and their application in land management. *Physiol. Ecol. Jpn.* **1990**, *27*, 131–146.

21. Haber, W. Using landscape ecology in planning and management. In *Changing Landscapes: An Ecological Perspective*; Springer: Berlin, Germany, 1990; pp. 217–232.

22. Hong, S.K.; Kim, S.; Cho, K.H.; Kim, J.E.; Kang, S.; Lee, D. Ecotope mapping for landscape ecological assessment of habitat and ecosystem. *Ecol. Res.* **2004**, *19*, 131–139. [CrossRef]

23. Shen, L.; Wu, L.; Dai, Y.; Qiao, W.; Wang, Y. Topic modelling for object-based unsupervised classification of VHR panchromatic satellite images based on multiscale image segmentation. *Remote Sens.* **2017**, *9*, 840. [CrossRef]

24. Nemmaoui, A.; Aguilar, M.; Aguilar, F.; Novelli, A.; García Lorca, A. Greenhouse crop identification from multi-temporal multi-sensor satellite imagery using object-based approach: A case study from Almería (Spain). *Remote Sens.* **2018**, *10*, 1751. [CrossRef]

25. Ruan, R.; Ren, L. Urban ecotope mapping using QuickBird imagery. In Proceedings of the IEEE International Geoscience and Remote Sensing Symposium, Barcelona, Spain, 23–28 July 2007; pp. 2963–2966.

26. Geerling, G.; Vreeken-Buijs, M.; Jesse, P.; Ragas, A.; Smits, A. Mapping river floodplain ecotopes by segmentation of spectral (CASI) and structural (LiDAR) remote sensing data. *River Res. Appl.* **2009**, *25*, 795–813. [CrossRef]

27. Janowski, L.; Trzcinska, K.; Tegowski, J.; Kruss, A.; Rucinska-Zjadacz, M.; Pocwiardowski, P. Nearshore Benthic Habitat Mapping Based on Multi-Frequency, Multibeam Echosounder Data Using a Combined Object-Based Approach: A Case Study from the Rowy Site in the Southern Baltic Sea. *Remote Sens.* **2018**, *10*, 1983. [CrossRef]

28. Guilbert, E.; Moulin, B. Towards a common framework for the identification of landforms on terrain models. *ISPRS Int. J. Geo-Inf.* **2017**, *6*, 12. [CrossRef]

29. Gerçek, D.; Toprak, V.; Strobl, J. Object-based classification of landforms based on their local geometry and geomorphometric context. *Int. J. Geogr. Inf. Sci.* **2011**, *25*, 1011–1023. [CrossRef]

30. Drăguţ, L.; Eisank, C. Automated object-based classification of topography from SRTM data. *Geomorphology* **2012**, *141*, 21–33. [CrossRef] [PubMed]

31. Gessler, P.; Chadwick, O.; Chamran, F.; Althouse, L.; Holmes, K. Modeling soil–landscape and ecosystem properties using terrain attributes. *Soil Sci. Soc. Am. J.* **2000**, *64*, 2046–2056. [CrossRef]

32. Kravchenko, A.N.; Bullock, D.G. Correlation of corn and soybean grain yield with topography and soil properties. *Agron. J.* **2000**, *92*, 75–83. [CrossRef]

33. Sternberg, M.; Shoshany, M. Influence of slope aspect on Mediterranean woody formations: Comparison of a semiarid and an arid site in Israel. *Ecol. Res.* **2001**, *16*, 335–345. [CrossRef]

34. Baatz, M.; Schäpe, A. Multiresolution Segmentation—An optimization approach for high quality multi-scale image segmentation. In *Angewandte Geographische Informationsverarbeitung XII*; Strobl, J., Blaschke, T., Griesebner, G., Eds.; Wichmann-Verlag: Heidelberg, Germany, 2000; pp. 12–23.

35. Radoux, J.; Bogaert, P. Accounting for the area of polygon sampling units for the prediction of primary accuracy assessment indices. *Remote Sens. Environ.* **2014**, *142*, 9–19. [CrossRef]

36. OpenStreetMap Contributors. Planet Dump. 2017. Available online: https://www.openstreetmap.org (accessed on 24 September 2018).

37. Hijmans, R.; Cameron, S.; Parra, J.; Jones, P.; Jarvis, A. Very high resolution interpolated climate surfaces for global land areas. *Int. J. Climatol.* **2005**, *25*, 1965–1978. [CrossRef]

38. Radoux, J.; Bogaert, P. Good Practices for Object-Based Accuracy Assessment. *Remote Sens.* **2017**, *9*, 646. [CrossRef]

39. Drăguţ, L.; Tiede, D.; Levick, S.R. ESP: A tool to estimate scale parameter for multiresolution image segmentation of remotely sensed data. *Int. J. Geogr. Inf. Sci.* **2010**, *24*, 859–871. [CrossRef]

40. Radoux, J.; Defourny, P. Quality assessment of segmentation results devoted to object-based classification. In *Object-Based Image Analysis*; Springer: Berlin, Germany, 2008; pp. 257–271.

Article

Edge Dependent Chinese Restaurant Process for Very High Resolution (VHR) Satellite Image Over-Segmentation

Hong Tang *, Xuejun Zhai and Wei Huang

Beijing Key Laboratory of Environmental Remote Sensing and Digital Cities, Faculty of Geographical Science, Beijing Normal University, Beijing 100875, China; zhai.xuejun@mail.bnu.edu.cn (X.Z.); huangwei@mail.bnu.edu.cn (W.H.)
* Correspondence: tanghong@bnu.edu.cn; Tel.: +86-10-5880-6401

Received: 15 August 2018; Accepted: 18 September 2018; Published: 21 September 2018

check for
updates

Abstract: Image over-segmentation aims to partition an image into spatially adjacent and spectrally homogeneous regions. It could reduce the complexity of image representation and enhance the efficiency of subsequent image processing. Previously, many methods for image over-segmentation have been proposed, but almost of them need to assign model parameters in advance, e.g., the number of segments. In this paper, a nonparametric clustering model is employed to the over-segmentation of Very High Resolution (VHR) satellite images, in which the number of segments can automatically be inferred from the observed data. The proposed model is called the Edge Dependent Chinese restaurant process (EDCRP), which extends the distance dependent Chinese restaurant process to make full use of local image structure information, i.e., edges. Experimental results show that the presented methods outperform state of the art methods for image over-segmentation in terms of both metrics based direct evaluation and classification based indirect evaluation.

Keywords: image over-segmentation; distance dependent Chinese restaurant process; nonparametric Bayesian clustering model; superpixels

---

1. Introduction

With the development of imaging techniques and the space satellite manufacturing, both spectral and spatial resolutions of Very High Resolution (VHR) satellite images have been improved significantly [1–3]. Consequently, VHR satellite images have been extensively applied to many applications, such as natural disaster monitoring [4,5], land cover change detection [6,7], environmental protection [8], and agricultural production [9], and so on [10,11]. Along with the increase in spatial resolution, it has becomes a considerable challenge for traditional methods to extraction information from VHR satellite images. Image over-segmentation is a common way to simplify image representations and speed image processing, by partitioning images into spatially adjacent and spectrally homogeneous regions [12,13].

Image over-segmentation has been widely used in many applications, including computer vision [14,15], object recognition [16,17], image retrieval [18,19], and image classification [20,21]. The following properties might be expected for a "good" over-segmentation result: (1) pixels within an segment should be spatially adjacent and have similar features; (2) the boundaries of segments should adhere well to the "meaningful" image boundaries; and, (3) each segment resides within only one real geo-object. Many over-segmentation algorithms have been used to segment everyday photos and pictures, and they aimed to achieve abovementioned properties, including the Simple Linear Iterative Clustering (SLIC) [22], Entropy Rate Superpixel segmentation (ERS) [23], normalized cut (NC) [24], edge-augmented mean-shift (ED) [25], and so on [26,27]. These methods can be roughly divided into four categories, including threshold-based algorithms [28,29], edge-based algorithms [30,31], graph-based algorithms [32,33], and clustering-based methods [34,35]. It is rather simple and efficient for the threshold-based algorithms to obtain over-segmentation from images. However, it is very difficult to choose the appropriate threshold. Although the boundary recall is often rather high in the results of edge-based methods, the results are often sensitive to the quality of the detected edges. It is straightforward for the graph-based algorithms to utilize the spatial constraints by measuring the affinity in terms of spectral or structural similarity.

Clustering-based methods are widely used for image over-segmentation by embedding spatial information. The SLIC is the most typical method to over-segmentation by clustering [22]. The Entropy Rate Superpixel (ERS) also treats the over-segmentation as a clustering problem [23]. The main difficulty for clustering-based methods to image segmentation is how to estimate the suitable number of clusters before clustering. As a Bayesian nonparametric clustering model, the Chinese restaurant process (CRP) mixture model [36], and its variants provide a principled way to infer the number of clusters from the observed data. An underlining assumption of these nonparametric clustering models is that the observed data in each group is modeled as an exchangeable sequence of random variables. In order words, for an exchangeable sequence of random variables, any permutation of the sequence has the same joint probability distribution as the original sequence [37]. If the CRP is directly applied to image clustering, the spatial dependency among pixels would be lost. Generally speaking, there exist three kinds of ways enhance spatial dependency among pixels when the CRP and its variants are applied to image analysis, i.e., preprocessing, poster-processing, and directly modeling the interdependence among neighboring variables [36]. For example, by using over-segmentation of VHR satellite images as a preprocessing, the CRP and its variants have been successfully applied to image or feature fusion [38], geo-object category detection [39], and unsupervised classification [40], and so on. However, it is difficult for humans to assign an appropriate number of over-segments, since a large number of geo-objects with different size and shapes are scattered in VHR satellite images without uniform spatial distribution. Majority voting using neighboring pixels are often used as a poster-processing to enhance spatial consistence of clustering labels [41]. However, the clustering results would be strongly dependent on both the presented neighboring system.

As an extension of the CRP, the distance dependent Chinese restaurant process (ddCRP) explicitly models the dependency between two random variables as a function of their distance [37]. The ddCRP was originally proposed for text modeling [37]. Whether and how can the ddCRP be effectively used for the over-segmentation of VHR satellite images? To answer this question, we systematically analyzed characteristics of the ddCRP. Specifically, we evaluated the impact of three components in the ddCRP model on the over-segmentation of VHR satellite images in this paper, i.e., the distance dependent term, the sampling probability term and the connection mode. Furthermore, we present an improved model for over-segmentation of VHR satellite images, which is termed as the Edge Dependent Chinese Restaurant Process (EDCRP).

The rest of this paper is organized as follows. In Section 2, we described the basic principle of the ddCRP, and analyzed the characteristics of its components. The EDCRP model was described in Section 3. Experimental results coupled with related discussions are given in Section 4. Some conclusions are drawn in Section 5.

2. Distance Dependent Chinese Restaurant Process

After the ddCRP is introduced in the first subsection, we systematically the characteristics of its three components, i.e., distance dependent term, likelihood term, and connection mode.

2.1. ddCRP

The CRP is a classical representation of the Dirichlet Process (DP) mixture model, which is an intuitional way to obtain a distribution over infinite partitions of the integers. Under the metaphor of the CRP, the random process is described, as follows. Customers enter a Chinese restaurant one by one for dinner. When a customer enters the restaurant, he or she chooses a table, which has been occupied by other customers who entered previously with a probability proportional to the number of customers already sitting at the table. Otherwise, one takes a seat at a new table with a probability proportional to a scale parameter α. The random process is given by

$$p(t_i = k | \mathbf{T}_{\neg i}, \alpha) \propto \begin{cases} n_k, & k < K \\ \alpha, & k = K + 1 \end{cases}, \tag{1}$$

where t_i represents the index of table chosen by the ith customer; $\mathbf{T}_{\neg i} = \{t_1, \ldots, t_{i-1}, t_{i+1}, \ldots, t_N\}$ denotes the table assignments of N customers with the exception of ith customer; K is the number of tables already occupied by some customers; and, n_k denotes the number of customers sitting at table k.

After all of the customers have taken their seats, a random partition has been realized. Customers sitting at the same table belong to a same cluster. This specific clustering property indicates a powerful probabilistic clustering method. The most important advantage of this algorithm is that the number of partitions does not need to be assigned in advance. The reason is that the CRP treats the number of clusters as a random variable that can be inferred from the observations by marginalizing out the random measure function.

Please note that, although the customers are described as entering the restaurant one by one in order, any permutation of their ordering has the same joint probability distribution of the original sequence. This is the exchangeability property of CRP mixture model. In practice, exchangeability is not a reasonable assumption for many realistic applications. As for image over-segmentation, it is necessary to identify the spatially contiguous segments where the same label is allocated to the neighbor pixels with homogeneous spectra. Therefore, the traditional CRP model is not suitable for image over-segmentation. To introduce the dependency among random variables, Blei et al. proposed the distance dependent Chinese restaurant process (ddCRP) in [37]. The partition, formed by assigning the customers to tables in the CRP, is replaced with the relationship between one customer and other customers in the ddCRP. Under the similar metaphor of the CRP, the ddCRP model states that each customer selects another customer to sit with, who has already entered the restaurant. Given the relationship among customers, it is easy to infer which customers will sit at the same table. Therefore, the tables are a byproduct of the customer assignments. Let f denote the decay function, d_{ij} the distance measure between two observations i and j, α the scaling parameter, and c_i the customer assignment of ith customer, i.e., ith customer choose to sit with c_i customer. The random process is to allocate customer assignments according to

$$p(c_i = j) \propto \begin{cases} f(d_{ij}), & i \neq j \\ \alpha, & i = j \end{cases}, \tag{2}$$

where the distance measure d_{ij} is used to evaluate the difference between the two observations i and j, and the decay function $f(x)$ mediates how the distance affects the probability of connecting the two customers together. The farther the distance between the two data points, the smaller their probability of connecting with each other. The traditional CRP is a particular case of ddCRP when the decay function is assumed to be a constant, i.e., $f(d_{ij}) = 1$.

To bridge the gap of terminologies between feature clustering and image over-segmentation, Figure 1 is employed to construct the analogy between the metaphor of ddCRP and image over-segmentation. An image with 16 pixels is shown in Figure 1a, where each numbered square denotes a pixel. The image is a restaurant, which accommodates 16 customers under the metaphor of the ddCRP. Figure 1b shows a middle state during the random process of customer assignments, where each arrow indicates a customer choose to sit with another customer. All of customers, who chose to sit with, naturally form a cluster. In other words, they sit at a same table in the ddCRP.

This is illustrated in Figure 1c, where the pixels with a same color belong to a cluster. Under the terminology of image over-segmentation, a segment also consists of all of the pixels with a same color, where every two pixels within a segment can reach to each other along the inferred customer assignments in the ddCRP. Therefore, the inferred customer assignments are the key to derive the results of image over-segmentation.

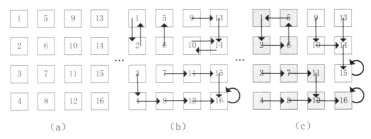

(a)   (b)   (c)

Figure 1. The illustration of distance dependent Chinese restaurant process (ddCRP) for image over-segmentation; (**a**) an image (i.e., a restaurant) with 16 pixels (i.e., customers) where each numbered square denotes a pixel; (**b**) each arrow indicates a customer choose to sit with another customer; and (**c**) a segment consists of pixels with a same color and every two pixels within a segment can reach to each other along the inferred customer assignments.

Given an image with N pixels $\mathbf{X} = \{x_1, \ldots, x_N\}$, the image over-segmentation is to partition pixels into multiple regions with label $\mathbf{Z} = \{z_1, \ldots, z_N\}$. In the ddCRP, the label $\mathbf{Z}$ is a by-product of customer assignment $\mathbf{C}$, whose posterior is given by $p(\mathbf{C}|\mathbf{X}) = \frac{p(\mathbf{C})p(\mathbf{X}|\mathbf{C})}{\sum_{c_{1:N}} p(\mathbf{C})p(\mathbf{X}|\mathbf{C})}$. The customer assignment, e.g., c_i, is assumed to be dependent on the distance between two pixels, e.g., d_{ij} and is independent of other customer assignments. Therefore, the posterior of customer assignment is proportional to

$$p(\mathbf{C}|\mathbf{X}) \propto \left[\prod_{i=1}^{N} p(c_i|\alpha, d, f) \right] p(\mathbf{X}|\mathbf{C}, G), \tag{3}$$

where the likelihood of image pixels is given by $p(\mathbf{X}|\mathbf{C}, G)$. It is difficult to compute the posterior because the ddCRP places a prior over a huge number of customer assignments. A simple Gibbs sampling method can be used to approximate the posterior by inferring its conditional probability

$$p(c_i|\mathbf{C}_{\neg i}, x_{1:N}, \alpha, d, f, G) \propto \begin{cases} f(d_{ij}) & \text{, situation 1} \\ f(d_{ij}) \frac{p\left(x_{(z(\mathbf{C}_{\neg i})=k)\cup(z(\mathbf{C}_{\neg i})=l)}\big|G\right)}{p\left(x_{(z(\mathbf{C}_{\neg i})=k)}\big|G\right)p\left(x_{(z(\mathbf{C}_{\neg i})=l)}\big|G\right)} & \text{, situation 2} \\ \alpha & \text{, situation 3} \end{cases} \tag{4}$$

where $\mathbf{C}_{\neg i}$ represent the customer assignments of all of the pixels with exception of ith pixel, and $x_{(z(\mathbf{C}_{\neg i})=k)}$ denotes the pixels associate with kth segment derived from $\mathbf{C}_{\neg i}$; G is a prior of the parameter θ, which is utilized to generate observations using a probability $p(x|\theta)$. In this paper, the prior G and the probability distribution $p(\theta)$ are assumed to be Dirichlet process [36] and multinomial distribution, respectively. The parameter θ can be marginalized out, since they are conjugated. As a family of stochastic processes, the Dirichlet process can be seen as the infinite-dimensional generalization of the Dirichlet distribution.

As the shown in Equation (4), there are three possible situations to generate a new assignment c_i. In situation 1, the assignment c_i points to another pixel but do not make any change in the over-segmentation result. The situation 2 means that two different segments are merged into a new segment due to the customer assignment c_i. For situation 3, the assignment c_i points to itself in a probability proportional to α.

In the following, we analyze the characteristics of the three components in the ddCRP, i.e., the distance dependent term $f(d_{ij})$, the likelihood of the image $p(x|G)$, and the connection mode among the pixels.

2.2. Distance Dependent Term

As for image over-segmentation, the dependence between the pixels can be naturally embedded into the ddCRP by the use of a spatial distance measure and a decay function, for example, the Euclidean distance between pixels $d_{spatial} = \sqrt{(row_i - row_j)^2 + (col_i - col_j)^2}$, where (row, col) is pixel's location in an image. The decay function is used to mediate the effects of distance between pixels on the sampling probability. The decay function should satisfy the following nonrestrictive assumptions; it should (1) be nonincreasing and (2) take nonnegative finite values, and (3) $f(\infty) = 0$. There are several types of decay functions [37], such as the window decay function, the exponential decay function and the logistic decay function. If the distance satisfies the preference, the decay function returns a larger value; otherwise, it returned a smaller value.

Since over-segments consist of a set of pixels that are spatially adjacent and spectrally homogeneous, only the neighboring pixels are considered in this paper, i.e., pixels i and j with spatial distance $d_{spatial} \leq 1$. As shown in Figure 1, the possible assignments of the 11th pixel consist of 7th, 10th, 12th, and 15th pixels. Therefore, the information on neighboring spatial locations has been introduced by this setting. In order to take spectral information between neighboring pixels into account in this model, the spectral difference between neighboring pixels is also introduced in the ddCRP. Spectral distance can be represented by the difference $d_{spectral} = |x_i - x_j|$, where x_i is the DN value of i-th pixel. So, the spectral difference and the spatial distance are combined to the final distance measure, $d_{ij} = d_{spatial} + d_{spectral}$. The ddCRP model with spatial and spectral distance abbreviated to spatial-spectral ddCRP model. In our previse work [42], empirical experiments showed that the spatial-spectral ddCRP model exhibits a promising performance.

2.3. Likelihood Term

Let the prior and probability distribution are the Dirichlet distribution and the multinomial distribution, respectively. Given customer assignments $\mathbf{c}$, the likelihood of kth segment is

$$p\left(x_{z(\mathbf{c})=k}\big|G\right) = \frac{\tau(\sum_w \beta_w)}{\prod_w \tau(\beta_w)} \div \frac{\tau\left(\sum_w \beta_w + n\left(x_{z(\mathbf{c})=k}\right)\right)}{\prod_{w=1} \tau\left(\beta_w + n\left(x_{z(\mathbf{c})=k} = w\right)\right)}, \tag{5}$$

where β denotes the parameter of the base distribution H, which is initialized while using a uniform distribution. The $\tau(.)$ represents the gamma function. Each visual word is denoted by w. In this paper, the visual word is a digital number (DN) value of panchromatic images. $n\left(x_{z(\mathbf{c})=k}\right)$ denotes the number of pixels belonging to the kth segment. As shown in Equation (2), kth and lth over-segments could be merged into a new one in a probability proportional to the production of distance dependent term and the likelihood ratio $\frac{p\left(x_{(z(\mathbf{c}_{-i})=k)\cup(z(\mathbf{c}_{-i})=l)}\big|G\right)}{p\left(x_{(z(\mathbf{c}_{-i})=k)}\big|G\right)p\left(x_{(z(\mathbf{c}_{-i})=l)}\big|G\right)}$. For the sake of simplifying description, the likelihood ratio is called the merging probability of two over-segments in the following. To reveal the characteristics of the merging probability, we performed five simulated experiments, where a paired of Gaussian distributions are utilized to simulate DN values of pixels within kth and lth over-segments. Table 1 lists parameters of Gaussian distributions, i.e., mean μ and variance σ, used in the five experiments. For each experiment, two subfigures are drawn on each row of Figure 2, where the left subfigure shows the paired of Gaussian distributions, and the right one is the distribution of the merging probability over the numbers of pixels within the paired over-segments. Within the right subfigure, the z-axis is the merging probability and x-, y-axis is the number of simulated pixels within the two over-segments, respectively.

Table 1. Parameters of paired Gaussian distributions used in the five experiments.

	(a)	(b)	(c)	(d)	(e)
kth over-segment	$\mu = 45, \sigma = 0.1$	$\mu = 45, \sigma = 2$	$\mu = 50, \sigma = 2$	$\mu = 55, \sigma = 2$	$\mu = 58, \sigma = 2$
lth over-segment	$\mu = 60, \sigma = 0.1$	$\mu = 60, \sigma = 2$	$\mu = 60, \sigma = 2$	$\mu = 60, \sigma = 2$	$\mu = 60, \sigma = 2$

It can be seen from Figure 2 that the distributions of the merging probability exhibit a similar pattern in the first four experiments, i.e., from (a) to (d). The merging probability is always less than 1. That is to say the two over-segments incline to remain to be separated, since they are not similar in terms of the rather large difference between the two mean DN values. Therefore, during the inferring process in the ddCRP, the customer assignment is allocated with a lower probability. In other words, any paired of pixels from the two over-segments would connect with each other with a lower probability.

Furthermore, for a given number of pixels within one segment, the merging probability decreases with the increase of number of pixels within the other segment. The underlining reason is that the dissimilarity between the two over-segments can be furthermore verified with more and more observations. In other words, a reliable state of customer assignments would be expected only when the number of pixels within each segment is up to some level. As shown in Figure 2a–d, the merging probability would be lower than 0.1 and become stable when the number of pixels is larger than about eight. This observation motivate us to use a number of pixels instead of individual pixel as a descriptor of each pixel in the improved model in Section 3.

As shown in Figure 2d, explicit local fluctuations occur in the distribution of merging probability since the two segments become more similar in terms of mean DN values. It can be seen from Figure 2e that the merging probability of experiment (e) looks very different from the other four experiments. The merging probability is significantly larger than 1 for most cases. This results show that the two over-segments would incline to be merged when they are similar in terms of mean DN values. This argument could also be verified furthermore when the number of pixels become more and more. It can be seen from Figure 2 that, if two segments are generated from dissimilar distributions, the merging probability will be less than 1, otherwise larger than 1. This suggests that it would be a good choice to let the parameter α equal to 1 in the ddCRP.

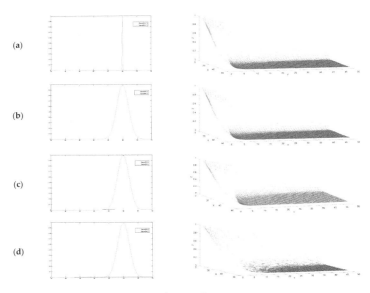

Figure 2. *Cont.*

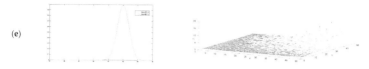

Figure 2. Five experiments are shown in subfigures (**a–e**), respectively. As for each experiment, the left subfigure shows a paired of Gaussian distributions, which are utilized to generate DN values of two over-segments, respectively; the right one is the distribution of the merging probability over the numbers of pixels within the two over-segments.

2.4. Connection Mode

In the ddCRP, the customer assignment is always allocated from some candidates that satisfy pre-specified constraints, which is called the connection mode in this paper. As shown in Figure 3a, the assignment of the 11th pixel is connected with one of its nearest neighbors (i.e., 7th, 10th, 12th, and 15th pixels) or itself according to the Gibbs sampling formula Equation (4). This setting implies a competition among the candidates and decreases the connecting probability of each candidate to some extent. Actually, it may be unnecessary to compare the connection relationship between pixels and other neighboring pixels. As shown in Figure 3b,c, each pixel might point to multiple pixels, i.e., multiple arrows. To discriminate different cases, one arrow that a pixel points to another pixel, is termed as "one-connection mode" (CM1), two arrows "two-connection mode" (CM2), and four arrows "four-connection mode" (CM4), respectively. Under the metaphor of CRP, the number of arrows indicates how many customers one could select to sit with during the inferring process.

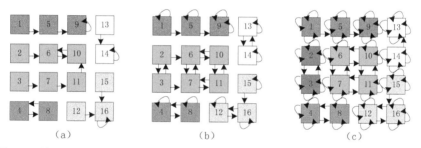

Figure 3. Three kinds of connection modes, (**a**) the one-connection mode (CM1), (**b**) the two-connection mode (CM2), and (**c**) the four-connection mode (CM4).

Figure 4 shows the results of over-segmentation over three kinds of geographic scenes, i.e., Suburban (S), Farmland (F), and Urban (U) areas, using the three type of connection modes. The suburban area contains sparse buildings, roads, and fields. The area of farmland consists of cultivated lands with similar shapes and slightly different spectra. In contrast, the urban area displays a large spectral variation, contains many buildings and roads, and has a complex structure. The images come from panchromatic TIANHUI satellites, which have a size of 200 × 200 pixels and a resolution of approximately 2 m.

Four metrics are employed to quantitatively evaluate the quality of image over-segmentations, i.e., the boundary recall (BR) [43], the achievable segmentation accuracy (ASA) [44], the under-segmentation error (UE) [45], and the degree of landscape fragmentation (LF) [42]. The BR is to measure how many percentage of real object boundaries have been discovered by an over-segmentation method. Based on the overlap between segments and real object regions, the ASA and UE is the percentage of pixels within or out of object regions, respectively. The last metric is to measure the degree of fragmentation for an over-segmentation result.

It can be seen from Table 2 that the CM1 is the best connection mode in terms of both BR and ASA for all of the three scenes. In contrast, the CM4 is the best one in terms of LF for all of the three scenes. The CM2 is in the middle among the three connection modes for all of the experiments with the exception of the UE of urban area. Please note that the metrics of both BR and ASA could increase with the number of segments. This also can reflected by the measurement of LF. The FL under CM1 is the highest among the three connection mode for all of three scenes. It also can be seen from Figure 4 that, under the CM1, there exist many segments of very small size, even many isolated points. In contrast, the size of segments is rather large under the CM4 and there explicitly exist under-segmented results. For example, many long and narrow objects, e.g., roads, are often incorrectly merged into larger regions under the CM4. To make full use of characteristics of different connection modes, both CM1 and CM2 will be utilized in our proposed method in Section 3.

Table 2. The quantitative evaluation of over-segmentations using the ddCRP over three type of geo-scenes, i.e., Suburban (S), Farmland (F), and Urban (U) areas, under three kinds of connection modes, i.e., one-connection mode (CM1), two-connection mode (CM2), and four-connection mode (CM4), respectively. For every evaluation metric, the best performance among different connection modes was marked as red or blue bold.

	Suburban			Farmland			Urban		
	CM1	CM2	CM4	CM1	CM2	CM4	CM1	CM2	CM4
BR	0.8	0.67	0.5	0.79	0.70	0.57	0.81	0.66	0.51
ASA	0.84	0.67	0.64	0.87	0.6	0.55	0.89	0.59	0.59
UE	0.07	0.09	0.15	0.11	0.12	0.15	0.09	0.07	0.09
LF	0.85	0.54	0.42	0.92	0.73	0.41	0.42	0.68	0.48

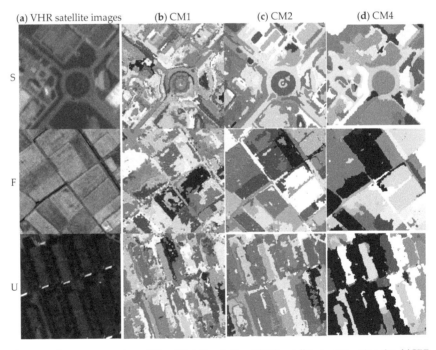

Figure 4. Over-segmentation of Very High Resolution (VHR) satellite images using the ddCRP with different connection modes: (**a**) VHR satellite images of three kinds of scenes, i.e., Suburb (S), Farmland (F), and Urban (U) areas, and corresponding over-segmentations using the ddCRP with (**b**) one-connection mode (CM1), (**c**) the two-connection mode (CM2), and (**d**) the four-connection mode (CM4), respectively.

3. Edge Dependent Chinese Restaurant Process

Based on the analysis about the ddCRP in Section 2, we present an improved method for the over-segmentation of VHR satellite images in this section, which is called Edge Dependent Chinese Restaurant Process (EDCRP). As shown in Figure 5, the EDCRP consists of three parts. The first part is edge detection from VHR satellite images. The second part is the selection of the feature descriptor and the connection mode that is based on these detected edges. The third part is image over-segmentation while using the spatial-spectral ddCRP.

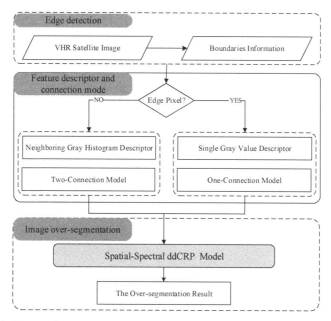

Figure 5. The flowchart of Edge Dependent Chinese Restaurant Process (EDCRP) for image over-segmentation.

3.1. Edge Detection

There exist many methods to extract edge information from images, e.g., the Sobel operator, the Roberts operator, the Prewitt operator, the Laplacian operator, or the Canny operator [46]. All of these methods can be utilized in the proposed method in order to improve the structure integrality of image over-segmentation. In this paper, the Canny operator was adopted since it works better by simultaneously considering both strong and weak edges.

Although the distance between neighboring pixels in the ddCRP could reflect the strength of a potential edge to some degree, there often exist many locations without meaningful edges, while the distance is rather large. Therefore, the detected edges could provide important prior knowledge of image structures for the ddCRP to work better. Given a pixel on an edge, the one-connection mode could provide an opportunity to motivate the model to choose a neighboring pixel with the shortest distance between them in the highest probability (under the assumption that the likelihood is the same value). Consequently, multiple over-segments would occur nearby the pixel in a higher probability. Meanwhile, if the one-connection mode is used everywhere, there would exist many segments of very small size, even many isolated points, as shown in Figure 4. Therefore, the two-connection mode coupled with a gray histogram descriptor is designed to remove these tiny over-segments, in particular isolated points.

3.2. Feature Descriptor and Connection Mode

As for each pixel, there exist two options for feature descriptors and connection modes, which is dependent on whether the pixel is on an edge or not. As for a pixel on an edge, its DN value is used as a feature descriptor and one-connection mode is used to decide which pixels could be tied together. A gray histogram based on neighboring pixels is used to describe a pixel, which is not on any edge and the two-connection mode is used. As shown in Figure 3, the gray histogram descriptor of 6-th pixel in Figure 3 will be constructed by the nine values of 1-th, 2-th, 3-th, 5-th, 6-th, 7-th, 9-th, 10-th, and 11-th pixels. The neighboring gray histogram descriptor is to remove the isolated points in the over-segmentation results. As discussed in the Section 2.3, the merging probability would be approximated to the scaling parameter α, when a segment has only a small number of pixels. As shown in Figure 2a–d, the merging probability would be lower than 0.1 and become stable when the number of pixels is larger than about 8. This observation motivate us to use nine pixels instead of individual pixel as a descriptor in the proposed model.

3.3. Spatial-Spectral ddCRP

As discussed in [42], the distance that was constructed with spatial and spectral features is better than only spatial distance. Therefore, both spatial and spectral distances are used in the proposed method. For the sake of clarity, it is called spatial-spectral ddCRP.

4. Experimental Results

In this section, we first described the experimental data. Then, we analyzed the effect of both feature descriptor and connection mode on image over-segmentations. Furthermore, the EDCRP is compared with state of the art methods in terms of quantitative evaluation matrices. At last, we discussed the efficiency and the possible extension of the EDCRP.

4.1. Experimental Data

As shown in Figure 6, two panchromatic images from different sensors are used in our experiments. The top-left image is a panchromatic TIANHUI image, which is in Mi Yun district of Beijing, China. The imaging time is 25 July 2013, the resolution is about 2 m, and the size of the image is 800×800 pixels. As shown in Figure 6a–d, four subareas of the TIANHUI image, coupled with real boundaries of interested geo-objects, will be employed to illustrate the qualitative effect of both the feature descriptor and the connection mode on over-segmentations in Section 4.2. The bottom-right image in Figure 6 is a panchromatic QUICKBIRD image, which was acquired on 22 April 2006. The area is located in Tong Zhou district of Beijing, China. The size of the image is 900×900 pixels with a resolution of 0.60 m.

4.2. Effect of Feature Descriptor and Connection Mode over Over-Segmentations

As shown in Figure 5, the EDCRP can be regard as a mixture of two kinds of feature descriptors and connection modes based on the spatial-spectral ddCRP model. In order to validate the necessary of this kinds of mixture, we compared the results of EDCRP with that of the spatial-spectral ddCRP model under two kinds of combination of both feature descriptor and connection-mode. Specifically, the first combination is that the DN value of each pixel is used as its feature and the one-connection is used. The second combination is that the feature descriptor of each pixel is the gray histogram of its neighboring pixels and the two-connection mode is used. For the sake of simplifying the notation, the two combinations are denoted by ddCRP_1 and ddCRP_2, respectively.

Only the TIANHUI image is used for the analysis in this section. Specifically, we first analyzed the over-segmentations of the two kinds of different combinations from the viewpoint of qualitatively visual inspection. Furthermore, we compared them with that of the proposed method. At last, four quantitative metrics are employed to measure the quality of these over-segmentations.

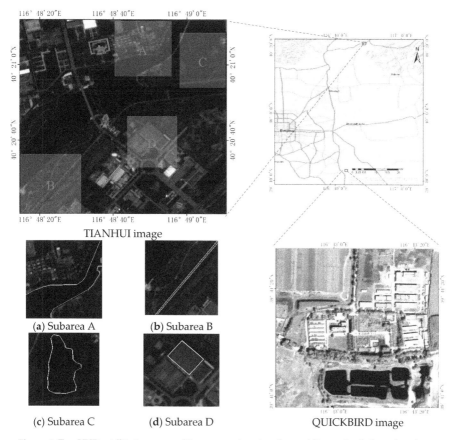

Figure 6. Two VHR satellite images used in our experiments, where subfigures (**a**–**d**) show the subarea
A, B, C, and D coupled with boundaries of interested geo-objects, respectively.

4.2.1. ddCRP_1

Figure 7 shows the boundaries of over-segments by using the ddCRP_1. Generally speaking,
there exist many tiny segments, even isolated points. As shown in the two subfigures, individual
geo-objects, e.g., water and building, are over-segmented into many segments. In addition, most of
the over-segments wriggle their way along the real boundaries of geo-objects. The structure of
over-segments is not very good. For example, there exist two edges along the top-right boundary of
the building in subarea D of Figure 7.

4.2.2. ddCRP_2

Figure 8 shows the boundaries of over-segmentation using the ddCRP_2. It can be seen that the
ddCRP_2 significantly outperforms over the ddCRP_1 in term of structure of over-segment's boudary.
On the one hand, from the viewpoint of overall visual inspection, boudaries of over-segments shown
in Figure 8a–d correspond well to boudaries of real geo-objects. For example, as shown in Figure 8c,d,
respectively, the boundaries of both water and building look very similar to the real one in terms of the
shape of geo-ojects. On the other hand, over-segments of the ddCRP_2 have been pushed into two
different directions relative to that of the ddCRP_1. As for the homogeneous regons, over-segments
become bigger. Heterogeneous regions have been partioned into more segments of ralative small
size. These two kinds of situation can be simultaneously seen in Figure 8d. The top-right part of the

building have been merged into a rather large segment, and its bottom-left part have been segmented into multiple small segments with regular boundaries.

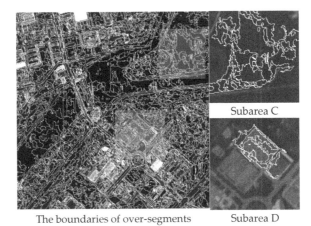

Figure 7. The boundaries of over-segments obtained by using the ddCRP_1 from the TIANHUI image. Both subareas C and D coupled with over-segments are shown in the two right-side subfigures.

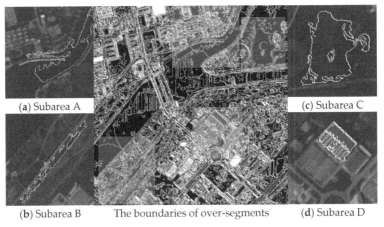

Figure 8. The boundaries of over-segments obtained by using the ddCRP_2 from the experimental image. The subfigures (**a**–**d**) are zoomed results from subareas A, B, C, and D, respectively.

In terms of the weakness of the ddCRP_2, on the one hand, there exist too much tiny segments over heterogeneous regions. On the other hand, some different geo-objects have been merged into a large segment. For example, it can be seen from Figure 8a that both the water and part of the land have been merged into a segment. In Figure 8b, the road also be merged with the vegetation along the road into a segment. These objects in Figure 6a,b are expected to be separated by over-segmentation algorithms. However, the ddCRP_2 did not achieve the expected result.

4.2.3. EDCRP

The result of EDCRP is shown in Figure 9. The result of EDCRP is similar with that of the ddCRP_2 in terms of the "shape" of over-segments. We argue that it is reasonable, since the EDCRP can be regarded as a mixture of both the ddCRP_1 and ddCRP_2, which are choosen dependent on pixels on edges or not. Generally speaking, the number of pixels on edges is explicitly less than that of pixels out of edges. Therefore, the EDCRP inherits the strongth of the ddCRP_2, i.e., well struture of over-segments' boudaries. Fortunately, the EDCRP also rules out the weakness of the ddCRP by using the ddCRP_1 over pixels on detected edges. In other words, the EDCRP would keep two regions being seperated where the ponits have been identified as a point on edges. For example, two different geo-objects in Figure 9a,b have been well seperated into different segments. It is very different from that in Figure 8a,b, which have been merged into large segments.

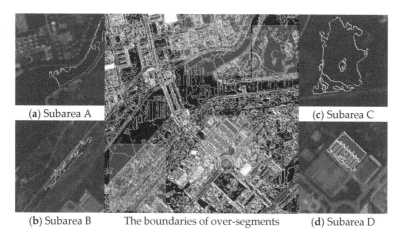

(a) Subarea A **(c) Subarea C**

(b) Subarea B   The boundaries of over-segments   **(d) Subarea D**

Figure 9. The boundaries of over-segments obtained by using the EDCRP from the experimental image. The subfigures (**a–d**) are zoomed results from subareas A, B, C, and D, respectively.

4.2.4. Quantitive Evaluation

Four metrics are adopted to quantitatively evaluate the quality of over-segmentations, i.e., BR, ASA, UE, and LF. As shown in Table 3, the EDCRP obtains the highest rate of boundary recall (i.e., BR) and achievable segmentation accuracy (ASA). Meanwhile, the EDCRP achieve the lowest error of under-segmentation (UE). As for landscape fragmentation (LF), the ddCRP_2 is the lowest one, with exception of the category building. This result can be explained by the observation in Figure 8 that there exist a large number of tiny segments. For a given image, both BR and ASA would often increase with the number of over-segments. In other words, both BR and ASA would increase with the decrease of LF for a given image. However, both BR and ASA of the EDCRP is the highest one, even when its LF is high than that of the ddCRP_2. In a word, the EDCRP outperforms both ddCRP_1 and ddCRP_2 in terms of all of the four quantitative metrics.

Table 3. The quantitative evaluation of over-segmentation.

	Building			Water			Bare Land			Road		
	ddCRP_1	ddCRP_2	EDCRP	ddCRP_1	ddCRP_2	EDCRP	ddCRP_1	ddCRP_2	EDCRP	ddCRP_1	ddCRP_2	EDCRP
BR	0.891	0.8946	0.9027	0.857	0.8436	0.8839	0.8405	0.8424	0.8918	0.9029	0.9030	0.9461
ASA	0.3834	0.4957	0.5023	0.1979	0.2013	0.2154	0.2057	0.3473	0.3619	0.2481	0.279	0.3619
UE	0.4698	0.4523	0.4016	0.3019	0.3024	0.2714	0.5467	0.5647	0.5183	0.4688	0.4636	0.4258
LF	0.0304	0.0214	0.0204	0.0595	0.0387	0.033	0.8405	0.0258	0.8918	0.0286	0.0231	0.0254

4.3. Comparative Experiments

In this subsection, the EDCRP is compared with three state-of-the-art methods, i.e., Turbo Pixels [47], SLIC [22], and ERS [23] while using the two panchromatic images with different resolutions. These methods are compared from the two viewpoints of both direct and indirect performance evaluation [48,49]. On the one hand, four metrics are utilized to directly evaluate the quality of over-segmentations, i.e., BR, ASA, UE, and LF. On the other hand, the quality of over-segmentations is indirectly evaluated in terms of image classification accuracy.

Both the SLIC and ERS achieve image over-segmentation by clustering. Turbo Pixels [47] uses a geometric-flow-based algorithm for acquiring an over-segmentation of an image. For the sake of fairly comparing, the number of clusters for other methods is assigned by the number of all over-segments inferred by the EDCRP. As shown in Figure 10, both SLIC and Turbo pixel generated over-segments with similar size and a more or less convex shape. Their boundaries do not exhibit image structure information. Over-segments produced by ERS have some structure information, but the result has higher complexity in flat area, especially in the water area.

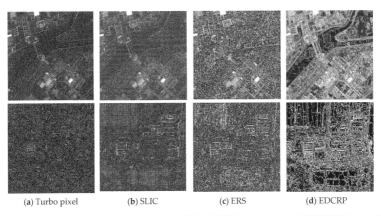

(a) Turbo pixel   (b) SLIC   (c) ERS   (d) EDCRP

Figure 10. The over-segmentation results over both TIANHUI (first row) and QUICKBIRD (second row) images using four over-segmentation methods, i.e., (**a**) Turbo pixel, (**b**) Simple Linear Iterative Clustering (SLIC), (**c**) Entropy Rate Superpixel segmentation (ERS), and (**d**) Edge Dependent Chinese Restaurant Process (EDCRP).

4.3.1. Metrics Based Direct Evaluation

Table 4 lists the values of four metrics of the three approaches for over-segmentation of the TIANHUI image. The ERS achieved the highest rate of boundary recall (i.e., BR) and achievable segmentation accuracy (ASA) among the three methods for all of four kinds of geo-object categories. As for the under-segmentation error (UE), the ERS is the lowest with the exception of the category water. The ERS also obtained a relative low value of LF. Based on these quantitative evaluation, the ERS outperform both the Turbo pixel and SLIC.

However, when compared with the EDCRP, the ERS achieved a lower value of both BR and ASA, and a rather higher value of both UE and LF. In other words, the EDCRP outperform all of the sate-of-the-art methods for over-segmentation in terms of four quantitative metrics. As for visual inspection, the over-segmentation result that is produced by ERS is similar with that of the EDCRP in terms of structure of over-segments. The significant difference is that ERS still produces rather small segments, even for homogeneous regions, e.g., water.

Table 4. The evaluation of over-segmentations of TIANHUI image while using Turbo Pixel (TP), SLIC and ERS.

	Building			Water			Bare Land			Road		
	TP	SLIC	ERS	TP	SLIC	ERS	TP	SLIC	ERS	TP	SLIC	ERS
BR	0.6675	0.6563	0.8723	0.625	0.6461	0.8327	0.6461	0.6229	0.8369	0.719	0.6847	0.8884
ASA	0.188	0.1766	0.2039	0.0112	0.0095	0.0141	0.0782	0.0715	0.0899	0.0956	0.0877	0.1009
UE	0.662	0.689	0.478	0.3247	0.3001	0.3197	0.6844	0.701	0.5772	0.5983	0.6226	0.4639
LF	0.0448	0.0372	0.0403	0.0672	0.0671	0.0648	0.0426	0.0385	0.0419	0.0482	0.0395	0.033

Table 5 lists the values of four metrics of the four approaches to the over-segmentation of the QUICKBIRD image. The EDCRP exhibits the best performance in terms of three metrics, i.e., BR, UE and LF. As for the achievable segmentation accuracy, EDCRP is only lower than the ERS, and higher than both Turbo pixel and SLIC.

Table 5. The quantitative evaluation of different over-segmentations of QUICKBIRD image.

Metric	Turbo Pixel	SLIC	ERS	EDCRP
BR	0.6953	0.7979	0.7675	0.8452
ASA	0.8912	0.8810	0.9191	0.8978
UE	0.0799	0.0679	0.0592	0.0542
LF	0.013	0.015	0.014	0.008

4.3.2. Classification Based Indirect Evaluation

Image over-segments are often used as objects for classification. In our experiments, random forests with 100 trees are trained as classifiers using randomly selected 50% over-segments. Table 6 lists both the Overall Accuracy (OA) and kappa of the classifier. The EDCRP outperforms other three approaches for image over-segmentation in terms of classification accuracy.

Table 6. The Overall Accuracy (OA) and kappa of the classifier over both the TIANHUI and QUICKBIRD images.

Image	Metric	Turbo Pixel	SLIC	ERS	EDCRP
TIANHUI	OA	0.7551	0.7706	0.7796	0.7878
	Kappa	0.6751	0.7012	0.7056	0.7152
QUICKBIRD	OA	0.8171	0.8116	0.7753	0.8189
	Kappa	0.7740	0.7681	0.7251	0.7759

Figure 11 shows the classification maps of the TIANHUI image. It can be seen from the classification maps that there exists rather explicit fingerprints of the over-segments. For example, the building in the subarea D, there exist misclassified blobs with convex shape in the subfigure (d) and (e) of Figure 11. In the subarea C, there exists explicit wriggled boundary of water in the subfigure (f) of Figure 11. In contrast, the classification map of EDCRP exhibits very good structure integrality for almost geo-objects. Specifically, as shown in the subarea B of Figure 11c, the wriggled *path* has been well classified.

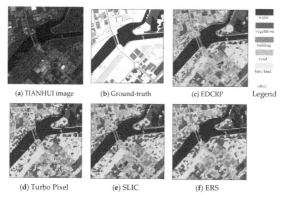

Figure 11. Classification maps of the TIANHUI image based on the over-segments; (**a**) is the TIANHUI image; (**b**) is the ground-truth of the image; (**c–f**) are the classification maps based on the over-segments produced by the EDCRP, Turbo Pixel, SLIC, and ERS, respectively. The legend shows the class labels in both the ground-truth and classification maps.

4.4. Discussions

In this section, we disscuss the efficiency and possible extensions of the EDCRP.

4.4.1. Efficiency

Since the number of clusters is also a random variable to be inferred in a nonparametric Bayesian clustering model, the process of clustering need more time than the parametric clustering method to be a convergence state. As shown in Equation (4), the posterior of customer assignments are approximated by using Gibbs sampling. A reliable sample would be generated only when the process of sampling reach a stable state. It can be seen from Table 7 that the efficiency of the proposed method is explicitly lower than that of the state of the art methods for image over-segmentation. Therefore, it deserves investigation as to how to significantly enhance the efficiency of the EDCRP in the future.

Table 7. The running time of different approaches for image over-segmentation.

	Turbo Pixel	SLIC	ERS	EDCRP
Time (s)	374	75	416	4981

4.4.2. Possible Extensions

Although the present model has only been applied to the VHR panchromatic image in this paper, it is straightforward to extend the EDCRP to the over-segmentation of multi-spectral images [50–52]. It can be seen from Equation (4) that the posterior of customer assignments can be approximated if both the likelihood and distance dependent terms can be well defined. Given a multispectral image, the likelihood in Equation (5) can be defined by replacing the multinomial distribution for discrete DN values with Gaussian distribution for multi-spectral satellite images.

5. Conclusions

In this paper, we systematically analyzed the characteristics of components in the ddCRP for VHR panchromatic image over-segmentation, i.e., distance dependent term, likelihood term, and connection mode among neighboring pixels. Furthermore, we present an improved nonparametric Bayesian method, which is called Edge Dependent Chinese Restaurant Process (EDCRP). Experimental results show that the presented methods outperform the state of the art over-segmentation methods in terms of both four evaluation metrics, i.e., under-segmentation error (UE), the boundary recall (BR), the achievable segmentation accuracy (ASA), and the degree of landscape fragmentation (LF) and classification accuracies. In the future, we would investigate how to speed the process of inferring customer assignments, and extend the present method to over-segment VHR multi-spectral satellite images.

Author Contributions: Conceptualization, H.T.; Funding acquisition, H.T.; Investigation, H.T. and X.-J.Z.; Methodology, X.-J.Z.; Software, X.-J.Z. and W.H.; Supervision, H.T.

Acknowledgments: This work was supported by the National Key R&D Program of China (No. 2017YFB0504104) and the National Natural Science Foundation of China (No. 41571334).

Conflicts of Interest: The authors declare no conflicts of interest.

References

1. Benediktsson, J.A.; Chanussot, J.; Moon, W.M. Very high-resolution remote sensing: Challenges and opportunities [point of view]. *Proc. IEEE* **2012**, *100*, 1907–1910. [CrossRef]
2. Bjorgo, E. Very high resolution satellites: A new source of information in humanitarian relief operations. *Bull. Assoc. Inf. Sci. Technol.* **1999**, *26*, 22–24. [CrossRef]

3.   Marchisio, G.; Pacifici, F.; Padwick, C. On the relative predictive value of the new spectral bands in the WorldWiew-2 sensor. In Proceedings of the 2010 IEEE International Geoscience and Remote Sensing Symposium, Honolulu, HI, USA, 25–30 July 2010; pp. 2723–2726. [CrossRef]

4.   Saito, K.; Spence, R.J.; Going, C.; Markus, M. Using high-resolution satellite images for post-earthquake building damage assessment: A study following the 26 January 2001 gujarat earthquake. *Earthq. Spectra* **2004**, *20*, 145–169. [CrossRef]

5.   Tralli, D.M.; Blom, R.G.; Zlotnicki, V.; Donnellan, A.; Evans, D.L. Satellite remote sensing of earthquake, volcano, flood, landslide and coastal inundation hazards. *ISPRS J. Photogramm. Remote Sens.* **2005**, *59*, 185–198. [CrossRef]

6.   Shalaby, A.; Tateishi, R. Remote sensing and gis for mapping and monitoring land cover and land-use changes in the northwestern coastal zone of egypt. *Appl. Geogr.* **2007**, *27*, 28–41. [CrossRef]

7.   Zhou, W.; Huang, G.; Troy, A.; Cadenasso, M. Object-based land cover classification of shaded areas in high spatial resolution imagery of urban areas: A comparison study. *Remote Sens. Environ.* **2009**, *113*, 1769–1777. [CrossRef]

8.   Atzberger, C. Advances in remote sensing of agriculture: Context description, existing operational monitoring systems and major information needs. *Remote Sens.* **2013**, *5*, 949–981. [CrossRef]

9.   Brown, J.C.; Jepson, W.E.; Kastens, J.H.; Wardlow, B.D.; Lomas, J.M.; Price, K. Multitemporal, moderate-spatial-resolution remote sensing of modern agricultural production and land modification in the brazilian amazon. *GISci. Remote Sens.* **2007**, *44*, 117–148. [CrossRef]

10.   Reinartz, P.; Müller, R.; Lehner, M.; Schroeder, M. Accuracy analysis for DSM and orthoimages derived from spot hrs stereo data using direct georeferencing. *ISPRS J. Photogramm. Remote Sens.* **2006**, *60*, 160–169. [CrossRef]

11.   Holland, D.; Boyd, D.; Marshall, P. Updating topographic mapping in great britain using imagery from high-resolution satellite sensors. *ISPRS J. Photogramm. Remote Sens.* **2006**, *60*, 212–223. [CrossRef]

12.   Tighe, J.; Lazebnik, S. Superparsing: Scalable nonparametric image parsing with superpixels. In Proceedings of the European Conference on Computer Vision, Crete, Greece, 5–11 September 2010; Springer: Berlin, Germany, 2010; pp. 352–365.

13.   Haralick, R.M.; Shapiro, L.G. Image segmentation techniques. *Comput. Vis. Graph. Image Process.* **1985**, *29*, 100–132. [CrossRef]

14.   Schalkoff, R.J. *Digital Image Processing and Computer Vision*; Wiley: New York, NY, USA, 1989; Volume 286.

15.   Forsyth, D.A.; Ponce, J. *Computer Vision: A Modern Approach*; Prentice Hall Professional Technical Reference: Upper Saddle River, NJ, USA, 2002.

16.   Van de Sande, K.E.; Uijlings, J.R.; Gevers, T.; Smeulders, A.W. Segmentation as selective search for object recognition. In Proceedings of the 2011 IEEE International Conference on Computer Vision (ICCV), Barcelona, Spain, 6–11 November 2011; IEEE: New York, NY, USA, 2011; pp. 1879–1886.

17.   Belongie, S.; Malik, J.; Puzicha, J. Shape matching and object recognition using shape contexts. *IEEE Trans. Pattern Anal. Mach. Intell.* **2002**, *24*, 509–522. [CrossRef]

18.   Belongie, S.; Carson, C.; Greenspan, H.; Malik, J. Color-and texture-based image segmentation using em and its application to content-based image retrieval. In Proceedings of the 1998 Sixth International Conference on Computer Vision, Bombay, India, 4–7 January 1998; IEEE: New York, NY, USA, 1998; pp. 675–682.

19.   Sural, S.; Qian, G.; Pramanik, S. Segmentation and histogram generation using the HSV color space for image retrieval. In Proceedings of the 2002 International Conference on Image Processing, Rochester, NY, USA, 22–25 September 2002; IEEE: New York, NY, USA, 2002; p. II.

20.   Carson, C.; Belongie, S.; Greenspan, H.; Malik, J. Blobworld: Image segmentation using expectation-maximization and its application to image querying. *IEEE Trans. Pattern Anal. Mach. Intell.* **2002**, *24*, 1026–1038. [CrossRef]

21.   Harchaoui, Z.; Bach, F. Image classification with segmentation graph kernels. In Proceedings of the IEEE Conference on Computer Vision and Pattern Recognition (CVPR'07), Boston, MA, USA, 7–12 June 2007; IEEE: New York, NY, USA, 2007; pp. 1–8.

22.   Achanta, R.; Shaji, A.; Smith, K.; Lucchi, A.; Fua, P.; Süsstrunk, S. Slic superpixels compared to state-of-the-art superpixel methods. *IEEE Trans. Pattern Anal. Mach. Intell.* **2012**, *34*, 2274–2282. [CrossRef] [PubMed]

23.   Liu, M.-Y.; Tuzel, O.; Ramalingam, S.; Chellappa, R. Entropy rate superpixel segmentation. In Proceedings of the 2011 IEEE Conference on Computer Vision and Pattern Recognition (CVPR), Colorado Springs, CO, USA, 20–25 June 2011; IEEE: New York, NY, USA, 2011; pp. 2097–2104.

24. Shi, J.; Malik, J. Normalized cuts and image segmentation. *IEEE Trans. Pattern Anal. Mach. Intell.* **2000**, *22*, 888–905.
25. Comaniciu, D.; Meer, P. Mean shift analysis and applications. In Proceedings of the Seventh IEEE International Conference on Computer Vision, Kerkyra, Greece, 20–27 September 1999; IEEE: New York, NY, USA, 1999; pp. 1197–1203.
26. Yuan, J.; Gleason, S.S.; Cheriyadat, A.M. Systematic benchmarking of aerial image segmentation. *IEEE Geosci. Remote Sens. Lett.* **2013**, *10*, 1527–1531. [CrossRef]
27. Yuan, J.; Wang, D.; Cheriyadat, A.M. Factorization-based texture segmentation. *IEEE Trans. Image Process.* **2015**, *24*, 3488–3497. [CrossRef] [PubMed]
28. Davis, L.S.; Rosenfeld, A.; Weszka, J.S. Region extraction by averaging and thresholding. *IEEE Trans. Syst. Man Cybern.* **1975**, *5*, 383–388. [CrossRef]
29. Kohler, R. A segmentation system based on thresholding. *Comput. Graph. Image Process.* **1981**, *15*, 319–338. [CrossRef]
30. Zadeh, L.A. Some reflections on soft computing, granular computing and their roles in the conception, design and utilization of information/intelligent systems. *Soft Comput.* **1998**, *2*, 23–25. [CrossRef]
31. Senthilkumaran, N.; Rajesh, R. A study on edge detection methods for image segmentation. In Proceedings of the International Conference on Mathematics and Computer Science (ICMCS-2009), Chennai, India, 5–6 January 2009; pp. 255–259.
32. Sinop, A.K.; Grady, L. A seeded image segmentation framework unifying graph cuts and random walker which yields a new algorithm. In Proceedings of the 2007 IEEE 11th International Conference on Computer Vision (ICCV 2007), Rio de Janeiro, Brazil, 14–21 October 2007; IEEE: New York, NY, USA, 2007; pp. 1–8.
33. Grady, L. Multilabel random walker image segmentation using prior models. In Proceedings of the IEEE Computer Society Conference on Computer Vision and Pattern Recognition (CVPR 2005), San Diego, CA, USA, 20–25 June 2005; IEEE: New York, NY, USA, 2005; pp. 763–770.
34. Comaniciu, D.; Meer, P. Mean shift: A robust approach toward feature space analysis. *IEEE Trans. Pattern Anal. Mach. Intell.* **2002**, *24*, 603–619. [CrossRef]
35. Purohit, P.; Joshi, R. A new efficient approach towards k-means clustering algorithm. *Int. J. Comput. Sci. Commun. Netw.* **2013**, *4*, 125–129.
36. Sammut, C.; Webb, G.I. *Encyclopedia of Machine Learning*; Springer Science & Business Media: Berlin, Germany, 2011.
37. Blei, D.M.; Frazier, P.I. Distance dependent chinese restaurant processes. *J. Mach. Learn. Res.* **2011**, *12*, 2461–2488.
38. Mao, T.; Tang, H.; Wu, J.; Jiang, W.; He, S.; Shu, Y. A Generalized Metaphor of Chinese Restaurant Franchise to Fusing Both Panchromatic and Multispectral Images for Unsupervised Classification. *IEEE Trans. Geosci. Remote Sens.* **2016**, *54*, 4594–4604. [CrossRef]
39. Li, S.; Tang, H.; He, S.; Shu, Y.; Mao, T.; Li, J.; Xu, Z. Unsupervised Detection of Earthquake-Triggered Roof-Holes From UAV Images Using Joint Color and Shape Features. *IEEE Geosci. Remote Sens. Lett.* **2015**, *12*, 1823–1827.
40. Shu, Y.; Tang, H.; Li, J.; Mao, T.; He, S.; Gong, A.; Chen, Y.; Du, H. Object-Based Unsupervised Classification of VHR Panchromatic Satellite Images by Combining the HDP and IBP on Multiple Scenes. *IEEE Trans. Geosci. Remote Sens.* **2015**, *53*, 6148–6162. [CrossRef]
41. Yi, W.; Tang, H.; Chen, Y. An Object-Oriented Semantic Clustering Algorithm for High-Resolution Remote Sensing Images Using the Aspect Model. *IEEE Geosci. Remote Sens. Lett.* **2011**, *8*, 522–526. [CrossRef]
42. Zhai, X.; Niu, X.; Tang, H.; Mao, T. Distance dependent chinese restaurant process for VHR satellite image oversegmentation. In Proceedings of the 2017 Joint Urban Remote Sensing Event (JURSE), Dubai, United Arab Emirates, 6–8 March 2017; IEEE: New York, NY, USA, 2017; pp. 1–4.
43. Ren, X.; Malik, J. Learning a Classification Model for Segmentation. In *Null*; IEEE: New York, NY, USA, 2003; p. 10.
44. Nowozin, S.; Gehler, P.V.; Lampert, C.H. On parameter learning in CRF-based approaches to object class image segmentation. In Proceedings of the European Conference on Computer Vision, Crete, Greece, 5–11 September 2010; Springer: Berlin, Germany, 2010; pp. 98–111.
45. Veksler, O.; Boykov, Y.; Mehrani, P. Superpixels and supervoxels in an energy optimization framework. In Proceedings of the European Conference on Computer Vision, Crete, Greece, 5–11 September 2010; Springer: Berlin, Germany, 2010; pp. 211–224.
46. Jain, R.; Kasturi, R.; Schunck, B.G. *Machine Vision*; McGraw-Hill: New York, NY, USA, 1995; Volume 5.

47. Levinshtein, A.; Stere, A.; Kutulakos, K.N.; Fleet, D.J.; Dickinson, S.J.; Siddiqi, K. Turbopixels: Fast superpixels using geometric flows. *IEEE Trans. Pattern Anal. Mach. Intell.* **2009**, *31*, 2290–2297. [CrossRef] [PubMed]

48. Stutz, D.; Alexander, H.; Bastian, L. Superpixels: An evaluation of the state-of-the-art. *Comput. Vis. Image Underst.* **2018**, *166*, 1–27. [CrossRef]

49. Giraud, R.; Ta, V.T.; Papadakis, N. Robust superpixels using color and contour features along linear path. *Comput. Vis. Image Underst.* **2018**, *170*, 1–13. [CrossRef]

50. Benedek, C.; Descombes, X.; Zerubia, J. Building development monitoring in multitemporal remotely sensed image pairs with stochastic birth-death dynamics. *IEEE Trans. Pattern Anal. Mach. Intell.* **2012**, *34*, 33–50. [CrossRef] [PubMed]

51. Grinias, I.; Panagiotakis, C.; Tziritas, G. MRF-based Segmentation and Unsupervised Classification for Building and Road Detection in Peri-urban Areas of High-resolution. *ISPRS J. Photogramm. Remote Sens.* **2016**, *122*, 45–166. [CrossRef]

52. The, J.W.; Jordan, M.I.; Beal, M.J. Hierarchical Dirichlet processes. *J. Am. Stat. Assoc.* **2006**, *101*, 1566–1581.

Article

A Robust Rule-Based Ensemble Framework Using Mean-Shift Segmentation for Hyperspectral Image Classification

Majid Shadman Roodposhti [1,*], Arko Lucieer [1], Asim Anees [2,3] and Brett A. Bryan [4]

[1] Discipline of Geography and Spatial Sciences, School of Technology, Environments and Design, University of Tasmania, Churchill Ave, Hobart, TAS 7005, Australia

[2] School of Engineering, University of Tasmania, Churchill Ave, Hobart, TAS 7005, Australia

[3] Data Scientist Group, ProCan, Children's Medical Research Institute, 214 Hawkesbury Road, Westmead, NSW 2145, Australia

[4] Centre for Integrative Ecology, School of Life and Environmental Sciences, Deakin University, 221 Burwood Hwy, Burwood, VIC 3125, Australia

* Correspondence: majid.shadman@utas.edu.au

Received: 10 August 2019; Accepted: 29 August 2019; Published: 1 September 2019

check for
updates

Abstract: This paper assesses the performance of DoTRules—a dictionary of trusted rules—as a supervised rule-based ensemble framework based on the mean-shift segmentation for hyperspectral image classification. The proposed ensemble framework consists of multiple rule sets with rules constructed based on different class frequencies and sequences of occurrences. Shannon entropy was derived for assessing the uncertainty of every rule and the subsequent filtering of unreliable rules. DoTRules is not only a transparent approach for image classification but also a tool to map rule uncertainty, where rule uncertainty assessment can be applied as an estimate of classification accuracy prior to image classification. In this research, the proposed image classification framework is implemented using three world reference hyperspectral image datasets. We found that the overall accuracy of classification using the proposed ensemble framework was superior to state-of-the-art ensemble algorithms, as well as two non-ensemble algorithms, at multiple training sample sizes. We believe DoTRules can be applied more generally to the classification of discrete data such as hyperspectral satellite imagery products.

Keywords: image classification; ensemble; mean-shift; entropy; uncertainty map

1. Introduction

Image classification is a vital tool for generating maps for environmental monitoring [1]. While for decades, multispectral imagery archives have been used to produce thematic maps, hyperspectral imagery is potentially a better option because of the higher spectral resolution. Hyperspectral images, which often contain more than 50 bands of continuous spectral information [2], can provide considerably more spatial and spectral information about the visible objects in their recorded field of view than multispectral imagery [3]. Because of the quality of information, hyperspectral images are widely used in applications such as precision agriculture [4], biotechnology [5], mineral exploration [6], and land-cover investigations [7]. These various types of applications have generated interest in hyperspectral image classification that has grown rapidly during the past two decades, with significant progress [8].

Up to now, many popular machine-learning algorithms have been applied in hyperspectral image classification. These include instance-based [9], regression [10], regularization [11], decision tree [12], probabilistic [13], reinforcement learning [14], dimensionality reduction [15], ensemble [16],

Bayesian [17], maximum margin [18], evolutionary [19], clustering [9], association rule learning [20], artificial neural network [12,21,22] and deep learning [23] methods (see Figure 1). Regardless of the classification performance, many of these algorithms act as black-boxes, resulting in a poor recognition of the classification structure and robustness owing to the high-dimensionality of the data [24,25].

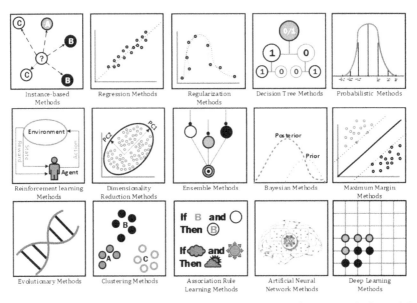

Figure 1. A visual illustration of different categories of machine learning methods used for image classification.

Recently, ensemble classification methods have received more attention from the machine learning community, resulting in their increased popularity in different applications such as hyperspectral image classification [26–28]. Nonetheless, as opposed to other black-box classification algorithms, rule-based ensembles have demonstrated the ability to inform the interpretation of classification schemes [29]. Rules are very general structures that offer an easily understandable and transparent way to find the most reliable class allocation [30]. The inferred logic of the model structure obtained by rule-based methods can be dissected, deciphered and applied out-of-the-box to new homogeneous classification problems. This is a major motivation, and it makes rule-based approaches more desirable compared with black-box approaches, even at the potential cost of a reduced classification accuracy [31]. This paper presents a simplified and novel rule-based ensemble framework based on the mean-shift and uncertainty assessment as a hyperspectral image classification tool, and we compare its performance against other state-of-the-art ensemble algorithms, where the mean-shift application is exclusive to the proposed framework. For the sake of simplicity throughout the paper, the proposed framework is referred to as DoTRules (Dictionary of Trusted Rules).

Here, we present DoTRules for hyperspectral image classification to provide a better and more transparent understanding of classification schemes, as well as accurate and robust classification performance. This adds to the growing literature of ensemble methods applied to the classification of hyperspectral data [32–34], especially those aimed at improving the performance of classification with acceptable clarity [35]. DoTRules is based on rules and uncertainty assessment. It was first introduced and applied to the calibration of land-use/cover change simulation models [36]. We assess the performance of the DoTRules algorithm as a novel rule-based classification framework modified to employ a bagging approach in order to boost accuracy. This accuracy boost is implemented by applying a thresholding assignment in order to extract trusted rules and then employing a novel voting approach to extract the class label recommended by the more trusted rules.

DoTRules extracts different subsets of training data from the full dataset, which can then be incorporated into boosting accuracy using a bagging approach designed to improve stability and accuracy. DoTRules has been found to perform well at modelling discrete data [37]. Since satellite imagery products inherently contain discrete digital numbers (DNs), DoTRules can work natively with them, quantifying the likelihood of belonging to a certain map class. It identifies classification rules and quantifies their frequencies so that some will be more influential than others. It also handles null values, which originate from unmatched rules between training and test samples. In addition, the uncertainty of every recognised classification rule is quantified using Shannon entropy. In simple terms, it scrutinises the uncertainty of each classification rule prior to assigning class labels based on their uncertainty value, so that the overall accuracy of classification can be improved. This not only results in boosting accuracy but also enables data analysts to spatially map every unique rule's uncertainty. In terms of applying DoTRules, every pixel of the target hyperspectral dataset corresponds to one rule from each rule set, and, after quantifying uncertainty, only the most competitive one is selected among all of the corresponding rules for a target pixel. Thus, as opposed to many other methods, DoTRules is not a black-box method, as the attributes and characteristics of every single rule can be openly observed. In addition, by quantifying the uncertainty of every rule we can then anticipate their hit ratio. This provides a tool for the spatial segregation of more reliable/accurate classified boundaries from less reliable/accurate ones prior to image classification.

The main objectives of this study are to: (1) demonstrate DoTRules as an accurate and transparent rule-based ensemble framework for hyperspectral image classification; (2) map the uncertainty of every unique classification rule as an estimate of the rules' hit ratio. This highlights the contribution of this paper, i.e., developing an accurate and transparent rule-based ensemble algorithm that provides a prior estimate of classification accuracy at the pixel level. Mapping the spatial distribution of classification accuracies is considered extremely beneficial for enhancing the capabilities of a classifier used as a land-use and land-cover map production tool based on satellite imagery [38,39]. Here, we describe the modified version of DoTRules for hyperspectral image classification, before demonstrating its application in three different study areas. We quantify the accuracy of DoTRules for hyperspectral image classification, and compare the results against some popular state-of-the-art ensemble approaches, i.e., extreme gradient boosting (XGBoost) [40,41], random forest (RF) [1,42–45], rotation forests (RoFs) [46–49], regularised random forest (RRF) [50,51], as well as two non-ensemble algorithms, namely, support vector machine (SVM) [52–56], and deep belief network (DBN) [57,58] as the classic deep learning method. Although SVM and DBN are not ensemble methods, they are included in our comparison because of their popularity, as they have been repeatedly used in recent hyperspectral image classification studies using Indian Pines, Salinas and Pavia University datasets. Finally, we discuss the advantages and disadvantages of the proposed approach for hyperspectral image classification.

2. Methods and Datasets

2.1. DoTRules

DoTRules is based on a dictionary of trusted rules. It is designed for prediction when a large amount of discrete data are involved. However, it may also be applied to continuous data after discretisation. This is similar to the RF [59,60] method insofar as rule sets are used to select the mode response (i.e., most frequent class label). However, instead of generating random trees, DoTRules operates by constructing many corresponding rules for every pixel (i.e., feature vector), which are derived from different rule sets. Each rule set is generated from a different combination of predictor variables in the ensemble run. For every unique rule, the most frequently occurring class label, which carries the highest probability of occurrence, is assigned [37,61]. However, as there are many rule sets, there may be many matching rules with defined class labels for a single data sample. To get the best (i.e., final) class label, a weighted majority filter (weighted mode) is applied on every available corresponding rule for a single data sample after the elimination of unreliable rules. The weighted majority filter puts more emphasis on those rules that are assembled by more components (i.e., matching variables) with less generalised class labels. The DoTRules procedure consists of the following steps implemented in R [62]:

STEP 1: *Segmentation analysis*

First, a data segmentation or segmentation analysis should be applied to each predictor variable $J=\{j_1,j_2, \dots, j_n\}$ before classification, where J is a defined set of *spectral bands/band combinations*, but not necessarily every spectral band or a possible combination. These homogeneous digital numbers (DNs) of the hyperspectral satellite image are then converted to segments. This is intended to partition m observations of the original image into S segments for each j in J, in which each DN in each segment (ideally) shares some common trait. Although various types of segmentation or even clustering algorithms can benefit the proposed classification framework, here we applied a *mean-shift* segmentation algorithm [63]. The mean-shift algorithm [63,64] is a recursive algorithm that allows us to execute a nonparametric mode-based segmentation. This is performed by a data segmentation based on a kernel density estimate of the probability density function associated with the data-generating process. The main motivation for applying a mean-shift algorithm is the fact that it is model-free and does not assume any prior distribution shape for data segments. Furthermore, it is robust to outliers and does not require a pre-specification of the number of segments.

In its standard form, the mean-shift algorithm works as follows. We observe a set of DN values from $x_1, \dots, x_m$, for each spectral band $J=\{j_1,j_2, \dots, j_n\}$. We fix a kernel function $ker f$ and a bandwidth parameter σ, and we apply the update rule:

$$x \leftarrow \frac{\sum_{i=1}^{m} ker f\left(\|\frac{x_i-x}{\sigma}\|\right)x_i}{\sum_{i=1}^{m} ker f\left(\|\frac{x_i-x}{\sigma}\|\right)} \tag{1}$$

where σ is a bandwidth parameter. The fundamental parameter in mean-shift algorithms is the bandwidth σ, which determines the number of segments [65]. Furthermore, regions with less than some pixel-count C may be optionally eliminated. To account for different spatial and spectral variances it is practical to choose a kernel window of size $\sigma = \sigma_s, \sigma_r$ with differing radii. σ_s is in the spatial domain, and σ_r is in the range domain. The statistics literature has developed various ways to estimate the bandwidth. One of them is the adaptive mean-shift where you let the bandwidth parameter vary for each data point. Here, the σ parameter is calculated using the kNN algorithm [66]. If $x_{i,S}$ is the k-nearest neighbour of x_i, then the bandwidth is calculated as:

$$\sigma_i = \|x_i - x_{i,S}\| \tag{2}$$

Here, the aim of the segmentation analysis is to summarise the input data and then minimise the required number of rules for correctly classifying pixels to their corresponding class label. As more accurate segments will improve the classification results, it is beneficial to apply the segmentation analysis on spectral band compositions composed of less similar spectral bands (i.e., within multidimensional space). Thus, a pairwise dissimilarity measure $dis(j_i, j_n)$ between spectral bands j_i and j_n, for $1 \leq i, j \leq n$ [67] can be applied to achieve more robust segments.

STEP 2: *Formatting the data*

In order to avoid mixing segment (S) values during the concatenation phase for the rule implementation in later steps, data segments should be formatted. Following the data segmentation, considering the maximum number of segments (S), the obtained data from step one should preferably be converted to two-digit (i.e., $S < 100$) or three-digit (i.e., $100 <= S < 1000$) numbers, or more. This is a requirement prior to the rule implementation. Hence, if a maximum value of S is under 100, the data should be formatted in a two-digit format (e.g.,3 = 03, 26 = 26), while if the maximum value of S is =>100 and < 1000, then the data should be in a three-digit format (e.g., 3 = 003, 26 = 026), and so forth.

STEP 3: *Splitting data into training and test samples*

Both our training and test sets will be in a tabular form, consisting of a set of pixels $I=\{i_1, i_2, ..., i_m\}$. Each pixel i in I has a value x_{ij} for each predictor variable J. Simply, x_{ij} is the converted segment value of the sample i in I and j in J. Thus, for each predictor variable j, x_{ij} can adopt one of a fixed set of possible values $\leq S$. Each pixel i has a corresponding class label $l_i \in L=\{l_1, l_2, \dots, l_h\}$, which are also discrete semantic attributes from the global set of class labels, such as corn, grass, oats, etc. It should be noted that to implement ensemble learners using DoTRules, we need to derive z sub-sets of our training dataset to construct different rule sets D containing individual classification rules d. This consists of all the available pixels in the primary training dataset but includes a different (random) combination of j in the feature vector.

STEP 4: *Creating a rule set*

For every z^{th} sub-set of the training set, we will concatenate values of a pixel x_{ij} for every j in J to form a rule set D. The concatenation of two or more characters is the string formed by them in a series (i.e., the concatenation of 001, 020, and 200 is 001020200). Equation (3) illustrates the pixel values for the segmented predictor variables concatenated for each pixel (row) i, thereby creating a rule for each pixel in the corresponding subset of the training dataset.

$$D_z = \begin{pmatrix} x_{11} \\ x_{21} \\ \vdots \\ x_{m1} \end{pmatrix} \| \begin{pmatrix} x_{12} \\ x_{22} \\ \vdots \\ x_{m2} \end{pmatrix} \| \cdots \| \begin{pmatrix} x_{1n} \\ x_{2n} \\ \vdots \\ x_{mn} \end{pmatrix} = \begin{bmatrix} x_{11} & x_{12} & \cdots & x_{1n} \\ x_{21} & x_{22} & \cdots & x_{2n} \\ \vdots & \vdots & \vdots & \vdots \\ x_{m1} & x_{m2} & \cdots & x_{mn} \end{bmatrix} = \begin{bmatrix} d_1 \\ d_2 \\ \vdots \\ d_{mn} \end{bmatrix} \tag{3}$$

Note that following the concatenation and extraction of rules (Equation (3)), every rule within a specific rule set has maintained its single class label $l_i \in L$. We then aggregate duplicate rules where pixels have exactly the same values for all criteria, leaving an efficient new rule set of unique rules D'_z. The frequency of occurrence of all potential class labels $l_i \in L$ is then calculated for each unique rule d' in D'_z:

$$\begin{bmatrix} L_1 \\ L_2 \\ \vdots \\ L_v \end{bmatrix} \begin{array}{c} \rightarrow \\ \rightarrow \\ \vdots \\ \rightarrow \end{array} \begin{bmatrix} f(l_1) & f(l_2) & f(l_3) & \cdots & f(l_h) \\ f(l_1) & f(l_2) & f(l_2) & \cdots & f(l_h) \\ \vdots & \vdots & \vdots & \vdots & \vdots \\ f(l_1) & f(l_2) & f(l_3) & \cdots & f(l_h) \end{bmatrix} \tag{4}$$

where v is the number of unique rules in D_z. The class label from the set L with the highest frequency (i.e., the mode) is then assigned to each corresponding unique rule d'. The total number of rule sets $D = [1, \dots, z]$ and the number of components in each rule set (i.e., the length of a rule) is user-defined. Although the classification accuracy may increase by using more rule sets, it will be at the expense of the computation cost. In terms of rule length, the accuracy of classification may not increase necessarily by the implementation of longer rules, where longer rules with more conditional components from the J set will model the training data too well (i.e., overfitting), resulting in less generalised responses for estimations of class labels, and vice versa (i.e., underfitting). Overall, as the quantity of matching pixels in the test dataset is inversely proportional to the length of rules, the longer rules with more components are more specific with fewer matches, while the shorter rules with fewer components are more general with many matches in the test dataset.

To ensure a more accurate estimation, the default value of z is set to 100 rule sets. Then, to avoid overfitting and underfitting issues, the number of predictor variables (j) used in every rule d within a specific rule set (length of rules within a considered rule set) is defined by a random function with a lower and upper bound defined by the user. This random function is called once, before creating every single rule set, to define the number and combination of components within that rule set. As the optimal combination of predictor variables is unknown, random band selection helps reduce the potential for the overfitting of the classifier. In this way, rules with various length will be implemented. The lower (*min*) and upper bound (*max*) for the length of rules (λ) in each rule set $D = [1, \dots, z]$ is a positive natural number defined by:

$$\lambda \begin{cases} \max(\lambda) \leq n \\ \min(\lambda) > 0 \end{cases} \tag{5}$$

where n is the number of selected predictor variables in J set. The number of rule sets, min and max values of λ can be further optimised using cross-validation.

STEP 5: *Calculating and mapping rule entropy.*

The aim of this step is to assess the uncertainty value of each rule. In information theory, entropy is the quantitative measure of system disorder, instability and uncertainty, and may be used to forecast the trend of a specified system. Entropy indicates the expected amount of information contained [68]. Here, the entropy value of every unique rule d' from a rule set D_z' is calculated based on the frequencies of each possible class label (Equation (4)) as:

$$e_{d'} = -\sum_{i=1}^{h} P_{l_i} \log_2 P_{l_i} \tag{6}$$

where $e_{d'}$ is the entropy of the unique rule d' and $^P l_i$ is the probability value of the class label $l_i \in L$. Here, h is the number of class labels in L. The general idea is that for a given rule, which may cover one or many pixels, the greater the probability of a class membership for a given class label, the less the uncertainty associated with that class. This provides a quantitative estimate of uncertainty for every single rule within different rule sets prior to assigning class labels. These estimates of uncertainty values can be applied to both the spatial mapping of rule uncertainty in classification, and to eliminating those unreliable rules with a high entropy from different subsets and/or rule sets before combining votes. The spatial distribution of uncertainty is quantified by mapping the entropy of each unique rule back to the corresponding pixels. These estimates of uncertainties are extremely beneficial and can be considered even prior to assigning class labels to pixels. Every time that DoTRules is applied to a training data subset, a class label of the highest frequency is allocated, and the entropy of that rule is calculated.

STEP 6: *Eliminating unreliable rules within all rule sets.*

After assessing the uncertainty of each individual rule, unreliable rules (i.e., rules with a high entropy) should be eliminated to improve the quality of the voting outcome, which directly affects the classification accuracy. Thus, every such rule d' (in D'_z), for which the $e_{d'}$ is greater than the user-defined threshold, is eliminated. In our study, we specified that if $e_{d'}$ is > 0.3 for a rule, and its corresponding pixel's frequency is < α (to avoid randomness), then the rule is considered to be unreliable and is eliminated accordingly. α is calculated as follows, keeping the random chance for a resultant entropy value under 0.05%:

$$\alpha = Ceil\left(\frac{\ln(0.05)}{\ln(1/h)}\right) \quad for \quad h > 1 \tag{7}$$

where h is the number of class labels.

STEP 7: *Classifying the test dataset.*

Above, we described the process of creating DoTRules and allocating the most likely class label for each rule based on the frequency. In the same way, class labels can now be assigned for the study area using another subset of the primary training dataset (i.e., implementing more rule sets). Every time a new rule set is implemented, the same procedure is followed to establish rules for the test dataset. We then match each test data rule with its equivalent training rules in the DoTRules using a many-to-one matching algorithm and allocate the most likely class label to each test data rule. This will be repeated every time that a weak learner is being implemented from every single rule set.

STEP 8: *Handling null values*

There is always a possibility of encountering null records in the test dataset while using DoTRules. In this situation, new pixels in the test dataset present combinations of states for criteria not encountered in the training data, which may increase the *out of bag* error. Handling null values is inevitable for maintaining the classification accuracy, where in the proposed ensemble framework using mean-shift it is a combined procedure. First, all rules are sorted based on their similarity, then every single null value is assigned to the class label of its closest (i.e., most similar) rule, based on the alphanumeric similarity of the constructed rules. However, the influence of these rules in combining votes is minimised as they are characterised by null entropy values.

STEP 9: *Combining votes*

In order to fulfil the classification procedure, this step is used to assign a final label to each pixel. To combine votes of each set of learners, we first remove all unreliable rules (with low or null entropy records) within every rule set using a thresholding approach. Afterwards, a mode filter is applied to the resultant class labels coming from sets of corresponding rules for each pixel. This mode function not only considers the frequency of class labels, but also considers the length of a rule as a weighted function. Since a rule is formed by concatenating n number of predictor variables (j), a rule that contains more predictor variables as components therefore has a higher weight in the mode function. Nonetheless, if none of the recognised reliable rules, for a certain pixel in the test dataset, is matched by any corresponding rule from the various training rule sets (derived from subsets of the training data), then the mode function will be applied to the corresponding labels of unreliable rules explained in STEP 8 with the same mode function.

STEP 10: *Calculating and mapping the hit ratio*

Calculating and mapping the hit ratio helps to visualise the spatial distribution of the classification error. Similar to the entropy value, which is calculated for every unique rule based on the frequencies of each possible class label, we map the hit ratio of every unique rule in our combined results back to the original pixels. DoTRules is rule-based, where every unique rule d' from a rule set D'_z corresponds to one or many pixels; thus, we can calculate the classification hit ratio of those rules using Equation (8):

$$A_{d'} = \sum_{i=1}^{h} l_i^{+} \bigg/ \sum_{i=1}^{h} l_i \qquad (8)$$

Here, l_i^{+} is the sum of the correct classified labels.

2.2. Rule Uncertainty Threshold

In using DoTRules for the classification of hyperspectral imagery, the class label of a rule is also described by both its entropy value and the frequency of all potential class labels (Figure 2). Therefore, a rule can be considered reliable if its entropy is less than 0.3 bits, which is calculated at least for n potential class labels (frequency $> \alpha$). However, it is important to note that among the reliable rules coming from the various rule sets for a certain pixel, those with a longer concatenated string (rule) will have more impact in combining final votes. This is mostly due to the fact that they are composed of more variables but meet the same uncertainty threshold, and hence can make more robust predictions. In other words, longer rules have fewer pixels with a specific class label, while shorter rules have more pixels belonging to multiple class labels. The fewer the pixels shared between different rules, the more accurate the classification results will be.

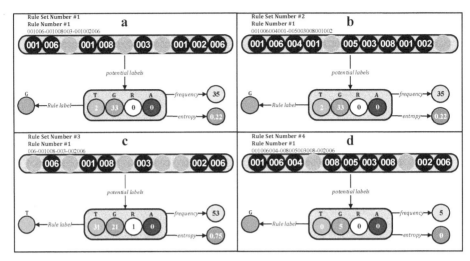

Figure 2. Schematic demonstration of (**a**,**b**) reliable and (**c**,**d**) unreliable rules extracted using DoTRules. The black circles represent the segment values of randomly selected spectral bands composing different rules for one target pixel. Considering rule sets number #1 and #2, the latter will have more impact in combining votes due to its larger length.

As the estimated entropy values for the distribution of response variables (class labels) with low frequencies are less reliable (Figure 2d) and may result from random chance, a second threshold is applied to the frequencies of potential labels. This will further improve the quality of the rule elimination process.

2.3. Comparing DoTRules with Other Methods

To measure and quantify DoTRules' performance, we implemented different classification algorithms, including XGBoost, RF, RoF, RRF, SVM, and DBN on the same datasets. These six algorithms are among the most popular methods for hyperspectral image classification, and they belong to three different categories of machine-learning methods. The first four algorithms are state-of-the-art ensemble methods, while SVM is a maximum margin classifier and DBN is a deep learning method. Thus, these methods provide appropriate benchmarks for assessing the performance of the DoTRules. XGBoost is an algorithm that has recently been dominating applied machine learning [69], and RF, RoF and RRF were selected because of both their natural similarity to DoTRules and performance in hyperspectral data classification [1,42–45]. They are also computationally efficient and suitable for large training datasets with many variables and can solve multiclass classification problems [70]. Furthermore, SVM [9,18,71] and DBN [57,58] algorithms have demonstrated promising results in previous studies. We compared the overall accuracy (OA) and kappa coefficient (*k*) of DoTRules with these various algorithms for hyperspectral image classification, using three different datasets from Indian Pines, Salinas and Pavia University (Figure 3).

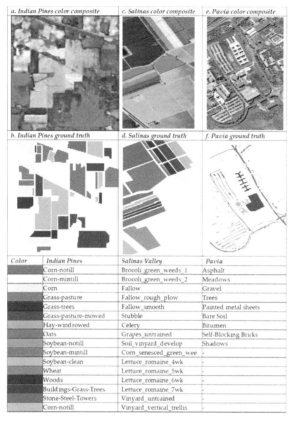

Color	Indian Pines	Salinas Valley	Pavia
	Corn-notill	Brocoli_green_weeds_1	Asphalt
	Corn-mintill	Brocoli_green_weeds_2	Meadows
	Corn	Fallow	Gravel
	Grass-pasture	Fallow_rough_plow	Trees
	Grass-trees	Fallow_smooth	Painted metal sheets
	Grass-pasture-mowed	Stubble	Bare Soil
	Hay-windrowed	Celery	Bitumen
	Oats	Grapes_untrained	Self-Blocking Bricks
	Soybean-notill	Soil_vinyard_develop	Shadows
	Soybean-mintill	Corn_senesced_green_wee	-
	Soybean-clean	Lettuce_romaine_4wk	-
	Wheat	Lettuce_romaine_5wk	-
	Woods	Lettuce_romaine_6wk	-
	Buildings-Grass-Trees	Lettuce_romaine_7wk	-
	Stone-Steel-Towers	Vinyard_untrained	-
	Corn-notill	Vinyard_vertical_trellis	-

Figure 3. False color composites and ground truth images of the datasets used to illustrate the image classification using DoTRules, including the (**a,b**) Indian Pines, (**c,d**) Salinas and (**e,f**) Pavia University datasets.

After tuning the required parameters of the above algorithms using the CARET package in R [72], a training process was implemented. In order to make a valid comparison, not only applicable to different study areas but also robust to variations of the portion of training and test sample sizes, different sample sizes of 1%, 5% and 10% were used. In addition, the overall accuracy value was taken as an average of five consecutive runs of each combination of algorithm and sample size. This was to avoid a sudden change in the overall accuracy value arising from changes in the training sample.

2.4. Datasets

DoTRules was tested using three hyperspectral image datasets (Figure 3), namely, the Indian Pines [22,73], Salinas [74,75] and Pavia University datasets [22,71]. Both the Indian Pines and Salinas datasets contain noisy bands due to water vapour, atmospheric effects, and sensor noise. All three datasets are available at http://www.ehu.eus/ccwintco/index.php?title=Hyperspectral_Remote_Sensing_Scenes. The mean spectral signatures of the three datasets is also demonstrated in Figure 4.

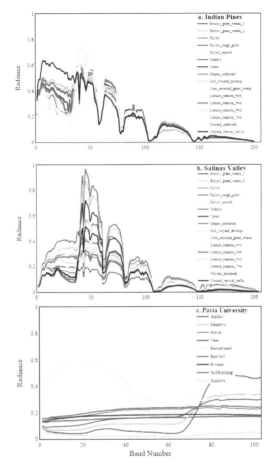

Figure 4. The mean spectral signatures of the (**a**) Indian Pines, (**b**) Salinas Valley and (**c**) Pavia University datasets.

The Indian Pines dataset is an AVIRIS image collected over the Indian Pines test site location, Indiana, USA. This dataset consists of 220 spectral bands in the same wavelength range as the Salinas dataset; however, four spectral bands are removed as they contain no data. This scene is a subset of a larger scene, and it contains 145 × 145 pixels covering 16 ground truth classes. We removed 20 spectral bands affected by water absorption and noise.

The Salinas image consists of 224 bands, and each band contains 512 × 217 pixels covering 16 classes. It was recorded by the AVIRIS sensor over Salinas Valley, CA, USA, with a spatial resolution of 3.7 m, and the spectral information ranging from 0.4 to 2.5 μm. We used 204 bands, after removing the water absorption bands.

The Pavia University dataset was collected by the Reflective Optics System Imaging Spectrometer (ROSIS) system that is a compact airborne imaging spectrometer. It consists of 103 spectral bands after removing the noisy bands, and 610 × 340 pixels for each band with a pixel resolution of 1.3 m. The ground truth image consists of nine classes.

3. Results

3.1. Simulation Experiments

For all three hyperspectral datasets, DoTRules was superior to all other algorithms in terms of the overall accuracy and kappa coefficient. However, considering the very low sample size (i.e., 1%) of the small-sized datasets (i.e., Indian Pines and Pavia University) it was not the most accurate approach. This is confirmed by the results of the accuracy assessment for the different sample sizes, which are averaged from five consecutive runs for a target sample size (Table 1).

Table 1. The accuracy assessment results of three applied datasets, including the overall accuracy (OA%) and kappa coefficient (κ) for all applied methods including support vector machine (SVM), deep belief network (DBN), extreme gradient boosting (XGBoost), random forest (RF), rotation forests (RoFs), regularised random forest (RRF), as well as Dictionary of Trusted Rules (DoTRules). The maximum values are highlighted in bold.

	Train	Test	SVM	DBN	XGboost	RF	RoF	RRF	DoTRules
Indian Pines	1%	50%	62.2	56.0	52.9	64.8	**70.5**	58.8	68.6
			0.558	0.486	0.453	0.593	**0.650**	0.521	0.640
	5%	50%	75.0	73.0	69.8	69.3	77.9	64.6	**87.3**
			0.708	0.689	0.656	0.644	0.725	0.588	**0.855**
	10%	50%	81.0	78.6	75.0	73.4	84.9	72.3	**93.2**
			0.781	0.755	0.710	0.693	0.788	0.675	**0.928**
Salinas	1%	50%	90.6	87.7	89.0	86.6	89.9	88.1	**91.5**
			0.895	0.862	0.877	0.850	0.881	0.867	**0.906**
	5%	50%	92.3	92.2	90.8	90.3	91.9	90.1	**97.2**
			0.914	0.913	0.898	0.892	0.908	0.888	**0.969**
	10%	50%	93.3	92.3	92.1	91.5	92.9	90.6	**98.7**
			0.925	0.914	0.912	0.905	0.918	0.895	**0.986**
Pavia	1%	50%	**92.0**	86.7	81.6	81.8	84.9	81.6	79.1
			0.893	0.820	0.748	0.749	0.790	0.732	0.720
	5%	50%	93.0	93.0	88.7	87.6	88.2	87.3	**93.1**
			0.907	0.906	0.849	0.833	0.871	0.817	**0.909**
	10%	50%	94.4	94.2	91.2	89.4	91.4	88.9	**96.2**
			0.925	0.920	0.882	0.857	0.895	0.850	**0.951**

The classification results also demonstrate that the DoTRules classification was able to closely match the spatial pattern of the ground truth image (Figure 5). These results were consistent across all three hyperspectral datasets. DoTRules was not only an accurate but also a transparent rule-based approach where the reliability (based on uncertainty) of each rule can be mapped. This is a desirable feature in remote sensing applications where the visual investigation of classification rules is informative.

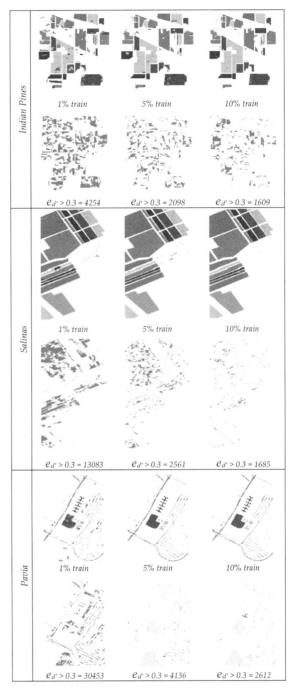

Figure 5. The DoTRules classification results and estimated pixel-based $e_{d'}$ for the Indian Pines, Salinas and Pavia datasets. The red pixels show the location of unreliable rules according to entropy thresholding ($e_{d'} > 0.3$, for $\alpha = 0.05$), while the grey pixels are reliable rules above the threshold. The red pixels are counted for each sample size.

3.2. Uncertainty Mapping

As DoTRules is rule-based, and each unique rule with its specific entropy value corresponds to one or more pixels, it is possible to estimate and map the uncertainty of every unique rule back to those pixels. This is a preliminary product of DoTRules, before assigning a class label to every pixel.

To illustrate the applicability of the entropy map to locate areas belonging to a low versus high classification accuracy, entropy values above and below the applied threshold ($e_d' = 0.3$, Equation (7)) were mapped to segregate regions which have more reliable and less reliable classification responses (Figure 5). In this way, the DoTRules spatial uncertainty map can facilitate a better understanding of uncertainty in the classified product and the segregation of more and less reliable geographic areas before assigning class labels to every pixel of the test sample dataset. This provides clear spatial insight into the uncertainty of the classification at an early stage of the classification process.

In developing and applying DoTRules, we have found that a larger sample size offers a higher classification accuracy where the number of less reliable rules with higher levels of uncertainty is reduced. Conversely, a smaller sample size, with fewer rules detected in our rule sets, was less able to capture the complexity of the hyperspectral image classification. This is mainly due to the fact that for DoTRules, training samples should be enough to cover all possible forms of rules. Figure 5 demonstrates the rule uncertainty for the Indian Pines, Salinas and Pavia University datasets using 1%, 5% and 10% training sample sizes.

3.3. Correspondence Between Uncertainty and Hit Ratio of Rules

In general, where there is low entropy (i.e., low uncertainty) for a rule within our rule set, the classification also tends to be more accurate. Simply, a lower entropy means there is just one clear answer (the mode class label) for a rule, while a high entropy indicates a more uniform distribution of the map class frequencies for that rule, which indicates a less reliable classification. Plotting hit ratio values against entropy values of every constructed rule among our various rule sets demonstrates that the hit ratio of rules can be defined by a polynomial function of their entropy value (Figure 6), which is supported by a strong coefficient of determination for all three datasets.

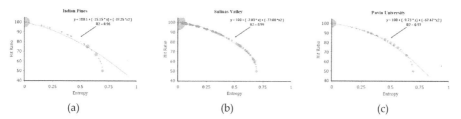

Figure 6. Entropy versus the hit ratio of rules for the (**a**) Indian Pines, (**b**) Salinas Valley and (**c**) Pavia University dataset 10% training sample sizes. The bubble sizes show the frequency of each rule among all corresponding rules from different rule sets before combining votes.

To further demonstrate the applicability of DoTRules' uncertainty product for the anticipation of the rule-exclusive hit ratio, we then applied the derived functions based on the correspondence of the hit ratio and entropy of the training data to predict the hit ratio of rules within the test datasets. The root mean square error (RMSE) values of the predicted hit ratios based on the entropy polynomial function were <1 for all three datasets (Table 2). Table 2 demonstrates that the uncertainty product of DoTRules may be applied to estimate the hit ratio of the classification rules in the context of the hyperspectral image classification.

Table 2. The prediction of rules' hit ratio based on the corresponding entropy values for a 10% training sample size.

Dataset	R	R-Squared	*p*-Value	Train RMSE	Test RMSE
Indian Pines	0.978	0.958	2.20×10^{-16}	0.3261	0.0972
Salinas Valley	0.996	0.993	2.20×10^{-16}	0.0195	0.1087
Pavia University	0.985	0.971	2.20×10^{-16}	0.0142	0.0628

4. Discussion

In this paper, we have presented a rule-based ensemble framework based on a mean-shift segmentation and uncertainty analysis, referred to as DoTRules (a Dictionary of Trusted Rules), for hyperspectral image classification. DoTRules constructs many rule sets composed of corresponding rules for each pixel in a hyperspectral image to predict the class of the test samples. When applied to different datasets and sample sizes, DoTRules proved to be an effective strategy for the classification of hyperspectral imagery, with promising results compared to other established algorithms. Furthermore, DoTRules enables both rule uncertainty and hit ratio mapping, which is an advantage for the users of classified land-use and land-cover maps created from remote sensing imagery. Below, we discuss improvements in hyperspectral image classification achieved using DoTRules.

4.1. The Overall Accuracy of Classification

According to our results, for all three applied hyperspectral datasets, the DoTRules ensemble framework was more accurate than the other applied classification algorithms for most training sample sizes (Table 1). This is due to the robust rule detection framework using mean-shift segmentation, where Shannon entropy is used to assess the uncertainty of individual rules for classification purpose. Here, the segmentation is done in a way where each DN in each segment (ideally) shares some common trait. This bears similarities with an object-oriented classification that involves the categorization of pixels based on the spatial relationship with the surrounding pixels. While pixel-based classification is exclusively based on the information in each pixel, object-based classification is based on information from a set of similar pixels (i.e., objects or image objects). Image objects are groups of pixels that are similar to one another based on the spectral properties (i.e., colour), size, shape, and texture, as well as context from a neighbourhood surrounding the pixels, in an attempt to mimic the type of analysis done by humans during visual interpretation. In addition, passing segment information to pixels and extracting reliable rules (i.e., low uncertainty rules) using minimum entropy through a voting system further preserves the high classification accuracy, especially when a representative training sample size is applied.

The observed increase in the overall accuracy of DoTRules' estimates when applying larger sample sizes may be due to an extra number of rules being detected and relatively fewer *null* records. Rules are very general structures that offer an easily understandable and transparent way to find the most reliable class allocation for a pixel [30]. As opposed to decision trees, every rule corresponds to only one pixel. This is unique to DoTRules and a common criticism of XGboost, RF, RoF, RRF and similar black-box algorithms [76,77]. Users can access all rules and their corresponding information, such as the rule ID, components of a rule (segment class for every selected band), true class label, probability (relative frequency) of every potential class label, rule entropy and hit ratio (accuracy) (Figure 2), while they are always connected to their corresponding pixels. This beneficial trait is highly valued in geoscience and remote sensing applications, especially in the context of land-use and land-cover mapping applications [38,78,79]. To be able to assign every pixel to a map class, each pixel should have at least one matching rule from various rule sets. Logically, the number of recognised rules within each individual rule set will be increased by a consequent increase in the training sample size (i.e., 1% to 10%), while the number of null records derived from unmatched rules between the test and training dataset will be reduced. Therefore, the greater the number of trusted rules, the greater the capability of our proposed framework to allocate test pixels into their true class labels.

4.2. Quantifying and Mapping the Uncertainty of Rules

While a few studies have successfully mapped the uncertainty of classification before image classification [38,80], one strength of DoTRules in hyperspectral image classification is its demonstrated ability to quantify the uncertainty of every identified transition rule using entropy values prior to the final classification (Figure 5). In other words, DoTRules was able to report the uncertainty of rules based on Shannon entropy, independent from the test dataset. The results from different hyperspectral datasets show that the lower the entropy value, the higher the hit ratio (Figure 6). Thus, considering the strong relationship between the entropy and hit ratio, it is possible to apply the entropy values as estimates of the hit ratio. The estimation of the rule uncertainty prior to the classification of a hyperspectral dataset aids in understanding the specific strengths and weaknesses of a classifier dealing with pixels containing a range of spectral information.

4.3. Quantifying Hit Ratio of Rules

DoTRules demonstrated the ability to quantify the rule-exclusive hit ratio using their corresponding entropy values (Table 2). Thus, the uncertainty product based on the entropy values can be applied to segregate areas of less and more reliable prediction independently of the test data availability. Thus, in the absence of a proper test dataset for the validation of classification results, rules' uncertainty values can be applied to represent their corresponding hit ratio. The collection of reliable ground truths for validation purposes is usually an expensive task in terms of time and economic costs [81]. Consequently, in many cases, it may not be possible to rely on test data to ensure good performance of a classifier. Accordingly, aside from using traditional accuracy metrics as a single number derived from a confusion matrix, mapping and thresholding the rule-exclusive hit ratio in a classification scheme is worthwhile for visualising general patterns of high and low accuracy values within the classified map and quantifying the accuracy of prediction in specific targeted locations.

4.4. Limitations of DoTRules and Future Work

Although the results obtained by DoTRules are encouraging, further comparative experiments with additional hyperspectral imagery datasets should be implemented. This can be more useful with a particular focus on assessing the classification performance at higher levels of disaggregation, such as a class-level accuracy assessment. As some of the required parameters for the DoTRules implementation are subjective, such as 1) the rule uncertainty threshold, 2) the minimum and maximum length of random rules and 3) the optimum number of rule sets, more research may be beneficial in the computational optimisation of DoTRules parameters. Our further work is focusing on the development of more computationally efficient schemes for the ensemble framework.

Another limitation of the proposed ensemble framework is the fact that the proposed framework is less efficient for very low sample sizes (i.e., 1% or less). DoTrules usually needs a larger training set to extract the underlying relationships between variables. This is a common requirement for all ensemble methods except RoF. Although RoF is the best performing algorithm for the 1% sample size of Indian Pines, it benefits from the transformation of the hyperspectral data.

5. Conclusions

We have applied DoTRules—a Dictionary of Trusted Rules—as an innovative ensemble framework for classifying hyperspectral data with high accuracy estimates compared with other popular classification algorithms. DoTRules' classification accuracy was superior to six other popular and state-of-the-art ensemble and non-ensemble algorithms. In the case of DoTRules, every rule within any rule set can be accessed, and their corresponding uncertainty value may be observed. This feature is unique to DoTRules and the absence of this ability underpins a common criticism of many ensemble algorithms (including many of the algorithms applied here) as black-box classifiers. Furthermore, DoTRules is also capable of quantifying and mapping the uncertainty of these classification rules, prior to the image classification where the uncertainty values can be applied as an estimate of the hit ratio. While the entropy product of DoTRules provides spatial insights, including the location of less reliable classification rules as well as more reliable ones, regardless of the test sample dataset availability, it can also certify and locate less accurate rules using the estimated hit ratio. The spatial exploration of rule uncertainty in hyperspectral image classification is beneficial for the early prediction of success or failure of a classifier in specific geographic locations. The uncertainty maps may also serve to enhance the application of map products by alerting map users to the spatial variation of rules' hit ratio over the entire mapped region. This, together with the simplicity and accuracy of DoTRules, indicates that the methodology offers new features and is ready for operational use by the remote sensing community.

Author Contributions: Conceptualization, M.S.R.; methodology, M.S.R.; validation, M.S.R. and A.A.; writing—original draft preparation, M.S.R., A.L. and B.A.B.; writing—review and editing, A.L., A.A. and B.A.B.; visualization, M.S.R.; supervision, A.L. and B.A.B.; funding acquisition, B.A.B. All authors read and approved the final manuscript.

Funding: This research was supported by CSIRO Australian Sustainable Agriculture Scholarship (ASAS) as a top-up scholarship to Majid Shadman Roodposhti, a PhD scholar, at the University of Tasmania (RT109121), School of Land and Food.

Acknowledgments: The authors greatly appreciate Maia Angelova Turkedjieva and Ye Zhu, Deakin University, for their suggestions to improve this manuscript. We also thank Monica Cuskelly, Associate Dean (Research), University of Tasmania, for editing the manuscript. We also thank the anonymous reviewers for their valuable comments and suggestions to improve the quality of the paper.

Conflicts of Interest: The authors declare no conflict of interest.

References

1. Chan, J.C.-W.; Paelinckx, D. Evaluation of random forest and adaboost tree-based ensemble classification and spectral band selection for ecotope mapping using airborne hyperspectral imagery. *Remote Sens. Environ.* **2008**, *112*, 2999–3011. [CrossRef]
2. Adep, R.N.; Vijayan, A.P.; Shetty, A.; Ramesh, H. Performance evaluation of hyperspectral classification algorithms on aviris mineral data. *Perspect. Sci.* **2016**, *8*, 722–726. [CrossRef]
3. Van der Meer, F.D.; van der Werff, H.M.A.; van Ruitenbeek, F.J.A.; Hecker, C.A.; Bakker, W.H.; Noomen, M.F.; van der Meijde, M.; Carranza, E.J.M.; Smeth, J.B.D.; Woldai, T. Multi- and hyperspectral geologic remote sensing: A review. *Int. J. Appl. Earth Obs. Geoinform.* **2012**, *14*, 112–128. [CrossRef]
4. Zarco-Tejada, P.J.; González-Dugo, M.V.; Fereres, E. Seasonal stability of chlorophyll fluorescence quantified from airborne hyperspectral imagery as an indicator of net photosynthesis in the context of precision agriculture. *Remote Sens. Environ.* **2016**, *179*, 89–103. [CrossRef]
5. Wakholi, C.; Kandpal, L.M.; Lee, H.; Bae, H.; Park, E.; Kim, M.S.; Mo, C.; Lee, W.-H.; Cho, B.-K. Rapid assessment of corn seed viability using short wave infrared line-scan hyperspectral imaging and chemometrics. *Sens. Actuators B Chem.* **2018**, *255*, 498–507. [CrossRef]
6. Rodger, A.; Laukamp, C.; Haest, M.; Cudahy, T. A simple quadratic method of absorption feature wavelength estimation in continuum removed spectra. *Remote Sens. Environ.* **2012**, *118*, 273–283. [CrossRef]
7. Chen, F.; Wang, K.; Van de Voorde, T.; Tang, T.F. Mapping urban land cover from high spatial resolution hyperspectral data: An approach based on simultaneously unmixing similar pixels with jointly sparse spectral mixture analysis. *Remote Sens. Environ.* **2017**, *196*, 324–342. [CrossRef]
8. Ma, X.; Wang, H.; Wang, J. Semisupervised classification for hyperspectral image based on multi-decision labeling and deep feature learning. *ISPRS J. Photogramm. Remote Sens.* **2016**, *120*, 99–107. [CrossRef]
9. Huang, K.; Li, S.; Kang, X.; Fang, L. Spectral–spatial hyperspectral image classification based on knn. *Sens. Imaging* **2015**, *17*, 1. [CrossRef]
10. Khodadadzadeh, M.; Li, J.; Plaza, A.; Bioucas-Dias, J.M. A subspace-based multinomial logistic regression for hyperspectral image classification. *IEEE Geosci. Remote Sens. Lett.* **2014**, *11*, 2105–2109. [CrossRef]
11. Peng, J.; Zhang, L.; Li, L. Regularized set-to-set distance metric learning for hyperspectral image classification. *Pattern Recognit. Lett.* **2016**, *83*, 143–151. [CrossRef]

12. Goel, P.K.; Prasher, S.O.; Patel, R.M.; Landry, J.A.; Bonnell, R.B.; Viau, A.A. Classification of hyperspectral data by decision trees and artificial neural networks to identify weed stress and nitrogen status of corn. *Comput. Electron. Agric.* **2003**, *39*, 67–93. [CrossRef]

13. Shao, Y.; Sang, N.; Gao, C.; Ma, L. Probabilistic class structure regularized sparse representation graph for semi-supervised hyperspectral image classification. *Pattern Recognit.* **2017**, *63*, 102–114. [CrossRef]

14. Zhong, Y.; Zhang, L. An adaptive artificial immune network for supervised classification of multi-/hyperspectral remote sensing imagery. *IEEE Trans. Geosci. Remote Sens.* **2012**, *50*, 894–909. [CrossRef]

15. Reshma, R.; Sowmya, V.; Soman, K.P. Dimensionality reduction using band selection technique for kernel based hyperspectral image classification. *Procedia Comput. Sci.* **2016**, *93*, 396–402. [CrossRef]

16. Naidoo, L.; Cho, M.A.; Mathieu, R.; Asner, G. Classification of savanna tree species, in the greater kruger national park region, by integrating hyperspectral and lidar data in a random forest data mining environment. *ISPRS J. Photogramm. Remote Sens.* **2012**, *69*, 167–179. [CrossRef]

17. Kayabol, K.; Kutluk, S. Bayesian classification of hyperspectral images using spatially-varying gaussian mixture model. *Digit. Signal Process.* **2016**, *59*, 106–114. [CrossRef]

18. Gao, L.; Li, J.; Khodadadzadeh, M.; Plaza, A.; Zhang, B.; He, Z.; Yan, H. Subspace-based support vector machines for hyperspectral image classification. *IEEE Geosci. Remote Sens. Lett.* **2015**, *12*, 349–353.

19. Feng, J.; Jiao, L.; Liu, F.; Sun, T.; Zhang, X. Unsupervised feature selection based on maximum information and minimum redundancy for hyperspectral images. *Pattern Recognit.* **2016**, *51*, 295–309. [CrossRef]

20. Guo, B.; Gunn, S.R.; Damper, R.I.; Nelson, J.D. Band selection for hyperspectral image classification using mutual information. *IEEE Geosci. Remote Sens. Lett.* **2006**, *3*, 522–526. [CrossRef]

21. Awad, M. Sea water chlorophyll-a estimation using hyperspectral images and supervised artificial neural network. *Ecol. Inform.* **2014**, *24*, 60–68. [CrossRef]

22. Yu, S.; Jia, S.; Xu, C. Convolutional neural networks for hyperspectral image classification. *Neurocomputing* **2017**, *219*, 88–98. [CrossRef]

23. Li, Y.; Xie, W.; Li, H. Hyperspectral image reconstruction by deep convolutional neural network for classification. *Pattern Recognit.* **2017**, *63*, 371–383. [CrossRef]

24. Castelvecchi, D. Can we open the black box of ai? *Nat. News* **2016**, *538*, 20. [CrossRef] [PubMed]

25. Goodfellow, I.J.; Erhan, D.; Carrier, P.L.; Courville, A.; Mirza, M.; Hamner, B.; Cukierski, W.; Tang, Y.; Thaler, D.; Lee, D.-H.; et al. Challenges in representation learning: A report on three machine learning contests. *Neural Netw.* **2015**, *64*, 59–63. [CrossRef] [PubMed]

26. Ayerdi, B.; Marqués, I.; Graña, M. Spatially regularized semisupervised ensembles of extreme learning machines for hyperspectral image segmentation. *Neurocomputing* **2015**, *149*, 373–386. [CrossRef]

27. Uslu, F.S.; Binol, H.; Ilarslan, M.; Bal, A. Improving svdd classification performance on hyperspectral images via correlation based ensemble technique. *Opt. Lasers Eng.* **2017**, *89*, 169–177. [CrossRef]

28. Ayerdi, B.; Graña, M. Hyperspectral image nonlinear unmixing and reconstruction by elm regression ensemble. *Neurocomputing* **2016**, *174*, 299–309. [CrossRef]

29. Tseng, M.-H.; Chen, S.-J.; Hwang, G.-H.; Shen, M.-Y. A genetic algorithm rule-based approach for land-cover classification. *ISPRS J. Photogramm. Remote Sens.* **2008**, *63*, 202–212. [CrossRef]

30. Russell, S.J.; Norvig, P.; Canny, J.F.; Malik, J.M.; Edwards, D.D. *Artificial Intelligence: A Modern Approach*; Prentice Hall: Upper Saddle River, NJ, USA, 2003; Volume 2.

31. Bauer, T.; Steinnocher, K. Per-parcel land use classification in urban areas applying a rule-based technique. *GeoBIT/GIS* **2001**, *6*, 24–27.

32. Benediktsson, J.A.; Garcia, X.C.; Waske, B.; Chanussot, J.; Sveinsson, J.R.; Fauvel, M. Ensemble methods for classification of hyperspectral data. In Proceedings of the IEEE International Geoscience and Remote Sensing Symposium (IGARSS 2008), Boston, MA, USA, 6–11 July 2008; IEEE: Piscataway, NJ, USA, 2008; pp. 62–65.

33. Ceamanos, X.; Waske, B.; Benediktsson, J.A.; Chanussot, J.; Fauvel, M.; Sveinsson, J.R. A classifier ensemble based on fusion of support vector machines for classifying hyperspectral data. *Int. J. Image Data Fusion* **2010**, *1*, 293–307. [CrossRef]

34. Xia, J.; Ghamisi, P.; Yokoya, N.; Iwasaki, A. Random forest ensembles and extended multiextinction profiles for hyperspectral image classification. *IEEE Trans. Geosci. Remote Sens.* **2018**, *56*, 202–216. [CrossRef]

35. Rokach, L. Ensemble-based classifiers. *Artif. Intell. Rev.* **2010**, *33*, 1–39. [CrossRef]

36. Shadman, M.; Aryal, J.; Bryan, B. Dotrules: A novel method for calibrating land-use/cover change models using a dictionary of trusted rules. In Proceedings of the MODSIM2017, 22nd International Congress

on Modelling and Simulation, Hobart, Australia, 3–8 December 2017; Syme, G., Hatton MacDonald, D., Fulton, B., Piantadosi, J., Eds.; Hobart, TAS, Australia, 2017; p. 508.

37. Roodposhti, M.S.; Aryal, J.; Bryan, B.A. A novel algorithm for calculating transition potential in cellular automata models of land-use/cover change. *Environ. Model. Softw.* **2019**, *112*, 70–81. [CrossRef]

38. Khatami, R.; Mountrakis, G.; Stehman, S.V. Mapping per-pixel predicted accuracy of classified remote sensing images. *Remote Sens. Environ.* **2017**, *191*, 156–167. [CrossRef]

39. Congalton, R.G. A review of assessing the accuracy of classifications of remotely sensed data. *Remote Sens. Environ.* **1991**, *37*, 35–46. [CrossRef]

40. Georganos, S.; Grippa, T.; Vanhuysse, S.; Lennert, M.; Shimoni, M.; Wolff, E. Very high resolution object-based land use-land cover urban classification using extreme gradient boosting. *IEEE Geosci. Remote Sens. Lett.* **2018**, *15*, 607–611. [CrossRef]

41. Loggenberg, K.; Strever, A.; Greyling, B.; Poona, N. Modelling water stress in a shiraz vineyard using hyperspectral imaging and machine learning. *Remote Sens.* **2018**, *10*, 202. [CrossRef]

42. Crawford, M.M.; Ham, J.; Chen, Y.; Ghosh, J. Random forests of binary hierarchical classifiers for analysis of hyperspectral data. In Proceedings of the 2003 IEEE Workshop on Advances in Techniques for Analysis of Remotely Sensed Data, Greenbelt, MD, USA, 27–28 October 2003; IEEE: Piscataway, NJ, USA, 2003; pp. 337–345.

43. Gislason, P.O.; Benediktsson, J.A.; Sveinsson, J.R. Random forests for land cover classification. *Pattern Recognit. Lett.* **2006**, *27*, 294–300. [CrossRef]

44. Ham, J.; Chen, Y.; Crawford, M.M.; Ghosh, J. Investigation of the random forest framework for classification of hyperspectral data. *IEEE Trans. Geosci. Remote Sens.* **2005**, *43*, 492–501. [CrossRef]

45. Lawrence, R.L.; Wood, S.D.; Sheley, R.L. Mapping invasive plants using hyperspectral imagery and breiman cutler classifications (randomforest). *Remote Sens. Environ.* **2006**, *100*, 356–362. [CrossRef]

46. Xia, J.; Du, P.; He, X.; Chanussot, J. Hyperspectral remote sensing image classification based on rotation forest. *IEEE Geosci. Remote Sens. Lett.* **2014**, *11*, 239–243. [CrossRef]

47. Xia, J.; Chanussot, J.; Du, P.; He, X. Spectral—Spatial classification for hyperspectral data using rotation forests with local feature extraction and markov random fields. *IEEE Trans. Geosci. Remote Sens.* **2015**, *53*, 2532–2546. [CrossRef]

48. Xia, J.; Falco, N.; Benediktsson, J.A.; Chanussot, J.; Du, P. Class-separation-based rotation forest for hyperspectral image classification. *IEEE Geosci. Remote Sens. Lett.* **2016**, *13*, 584–588. [CrossRef]

49. Feng, W.; Bao, W. Weight-based rotation forest for hyperspectral image classification. *IEEE Geosci. Remote Sens. Lett.* **2017**, *14*, 2167–2171. [CrossRef]

50. Izquierdo-Verdiguier, E.; Zurita-Milla, R.; Rolf, A. On the use of guided regularized random forests to identify crops in smallholder farm fields. In Proceedings of the 2017 9th International Workshop on the Analysis of Multitemporal Remote Sensing Images (MultiTemp), Brugge, Belgium, 27–29 June 2017; IEEE: Piscataway, NJ, USA, 2017; pp. 1–3.

51. Mureriwa, N.; Adam, E.; Sahu, A.; Tesfamichael, S. Examining the spectral separability of prosopis glandulosa from co-existent species using field spectral measurement and guided regularized random forest. *Remote Sens.* **2016**, *8*, 144. [CrossRef]

52. Fauvel, M.; Benediktsson, J.A.; Chanussot, J.; Sveinsson, J.R. Spectral and spatial classification of hyperspectral data using svms and morphological profiles. *IEEE Trans. Geosci. Remote Sens.* **2008**, *46*, 3804–3814. [CrossRef]

53. Tarabalka, Y.; Fauvel, M.; Chanussot, J.; Benediktsson, J.A. Svm-and mrf-based method for accurate classification of hyperspectral images. *IEEE Geosci. Remote Sens. Lett.* **2010**, *7*, 736–740. [CrossRef]

54. Bazi, Y.; Melgani, F. Toward an optimal svm classification system for hyperspectral remote sensing images. *IEEE Trans. Geosci. Remote Sens.* **2006**, *44*, 3374–3385. [CrossRef]

55. Cui, M.; Prasad, S. Class-dependent sparse representation classifier for robust hyperspectral image classification. *IEEE Trans. Geosci. Remote Sens.* **2015**, *53*, 2683–2695. [CrossRef]

56. Lv, Q.; Niu, X.; Dou, Y.; Wang, Y.; Xu, J.; Zhou, J. Hyperspectral image classification via kernel extreme learning machine using local receptive fields. In Proceedings of the 2016 IEEE International Conference on Image Processing (ICIP), Phoenix, AZ, USA, 25–28 September 2016; IEEE: Piscataway, NJ, USA, 2016; pp. 256–260.

57. Li, J.; Xi, B.; Li, Y.; Du, Q.; Wang, K. Hyperspectral classification based on texture feature enhancement and deep belief networks. *Remote Sens.* **2018**, *10*, 396. [CrossRef]

58. Chen, Y.; Zhao, X.; Jia, X. Spectral–spatial classification of hyperspectral data based on deep belief network. *IEEE J. Sel. Top. Appl. Earth Obs. Remote Sens.* **2015**, *8*, 2381–2392. [CrossRef]

59. Breiman, L. Bagging predictors. *Mach. Learn.* **1996**, *24*, 123–140. [CrossRef]
60. Breiman, L. Random forests. *Mach. Learn.* **2001**, *45*, 5–32. [CrossRef]
61. Roodposhti, M.S.; Aryal, J.; Pradhan, B. A novel rule-based approach in mapping landslide susceptibility. *Sensors* **2019**, *19*, 2274. [CrossRef]
62. R Core Team. *R: A Language and Environment for Statistical Computing*; R Foundation for Statistical Computing: Vienna, Austria, 2017.
63. Cheng, Y. Mean shift, mode seeking, and clustering. *IEEE Trans. Pattern Anal. Mach. Intell.* **1995**, *17*, 790–799. [CrossRef]
64. Fukunaga, K.; Hostetler, L. The estimation of the gradient of a density function, with applications in pattern recognition. *IEEE Trans. Inf. Theory* **1975**, *21*, 32–40. [CrossRef]
65. Silverman, B.W. *Density Estimation for Statistics and Data Analysis*; Routledge: Abingdon, UK, 2018.
66. Carreira-Perpinán, M.A. A review of mean-shift algorithms for clustering. *arXiv* **2015**, arXiv:1503.00687.
67. Huband, J.M.; Bezdek, J.C.; Hathaway, R.J. Bigvat: Visual assessment of cluster tendency for large data sets. *Pattern Recognit.* **2005**, *38*, 1875–1886. [CrossRef]
68. Shannon, C.E. A mathematical theory of communication. *ACM SIGMOBILE Mob. Comput. Commun. Rev.* **2001**, *5*, 3–55. [CrossRef]
69. Chen, T.; Guestrin, C. Xgboost: A scalable tree boosting system. In Proceedings of the 22nd ACM Sigkdd International Conference on Knowledge Discovery and Data Mining, San Francisco, CA, USA, 13–17 August 2016; ACM: New York, NY, USA, 2016; pp. 785–794.
70. Mahapatra, D. Analyzing training information from random forests for improved image segmentation. *IEEE Trans. Image Process.* **2014**, *23*, 1504–1512. [CrossRef] [PubMed]
71. Golipour, M.; Ghassemian, H.; Mirzapour, F. Integrating hierarchical segmentation maps with mrf prior for classification of hyperspectral images in a bayesian framework. *IEEE Trans. Geosci. Remote Sens.* **2016**, *54*, 805–816. [CrossRef]
72. Kuhn, M. Caret package. *J. Stat. Softw.* **2008**, *28*, 1–26.
73. Yang, C.; Tan, Y.; Bruzzone, L.; Lu, L.; Guan, R. Discriminative feature metric learning in the affinity propagation model for band selection in hyperspectral images. *Remote Sens.* **2017**, *9*, 782. [CrossRef]
74. Kianisarkaleh, A.; Ghassemian, H. Nonparametric feature extraction for classification of hyperspectral images with limited training samples. *ISPRS J. Photogramm. Remote Sens.* **2016**, *119*, 64–78. [CrossRef]
75. Luo, F.; Huang, H.; Duan, Y.; Liu, J.; Liao, Y. Local geometric structure feature for dimensionality reduction of hyperspectral imagery. *Remote Sens.* **2017**, *9*, 790. [CrossRef]
76. Palczewska, A.; Palczewski, J.; Robinson, R.M.; Neagu, D. Interpreting random forest models using a feature contribution method. In Proceedings of the 2013 IEEE 14th International Conference on Information Reuse and Integration (IRI), San Francisco, CA, USA, 14–16 August 2013; IEEE: Piscataway, NJ, USA, 2013; pp. 112–119.
77. Gislason, P.O.; Benediktsson, J.A.; Sveinsson, J.R. Random forest classification of multisource remote sensing and geographic data. In Proceedings of the 2004 IEEE International Geoscience and Remote Sensing Symposium, Anchorage, AK, USA, 20–24 September 2004; IEEE: Piscataway, NJ, USA, 2004; pp. 1049–1052.
78. Yang, X.; Chen, L.; Li, Y.; Xi, W.; Chen, L. Rule-based land use/land cover classification in coastal areas using seasonal remote sensing imagery: A case study from Lianyungang city, China. *Environ. Monit. Assess.* **2015**, *187*, 449. [CrossRef] [PubMed]
79. Lucas, R.; Rowlands, A.; Brown, A.; Keyworth, S.; Bunting, P. Rule-based classification of multi-temporal satellite imagery for habitat and agricultural land cover mapping. *ISPRS J. Photogramm. Remote Sens.* **2007**, *62*, 165–185. [CrossRef]
80. Bryan, B.A.; Barry, S.; Marvanek, S. Agricultural commodity mapping for land use change assessment and environmental management: An application in the Murray–darling basin, Australia. *J. Land Use Sci.* **2009**, *4*, 131–155. [CrossRef]
81. Bruzzone, L.; Prieto, D.F. Unsupervised retraining of a maximum likelihood classifier for the analysis of multitemporal remote sensing images. *IEEE Trans. Geosci. Remote Sens.* **2001**, *39*, 456–460. [CrossRef]

Article

Operational Large-Scale Segmentation of Imagery Based on Iterative Elimination

James D. Shepherd [1,*], Pete Bunting [2] and John R. Dymond [1]

[1] Landcare Research, Private Bag 11052, Palmerston North 4442, New Zealand; dymondj@landcareresearch.co.nz

[2] Department of Geography and Earth Sciences, Aberystwyth University, Aberystwyth SY23 3DB, UK; pete.bunting@aber.ac.uk

* Correspondence: shepherdj@landcareresearch.co.nz

Received: 28 February 2019; Accepted: 15 March 2019; Published: 18 March 2019

check for
updates

Abstract: Image classification and interpretation are greatly aided through the use of image segmentation. Within the field of environmental remote sensing, image segmentation aims to identify regions of unique or dominant ground cover from their attributes such as spectral signature, texture and context. However, many approaches are not scalable for national mapping programmes due to limits in the size of images that can be processed. Therefore, we present a scalable segmentation algorithm, which is seeded using k-means and provides support for a minimum mapping unit through an innovative iterative elimination process. The algorithm has also been demonstrated for the segmentation of time series datasets capturing both the intra-image variation and change regions. The quality of the segmentation results was assessed by comparison with reference segments along with statistics on the inter- and intra-segment spectral variation. The technique is computationally scalable and is being actively used within the national land cover mapping programme for New Zealand. Additionally, 30-m continental mosaics of Landsat and ALOS-PALSAR have been segmented for Australia in support of national forest height and cover mapping. The algorithm has also been made freely available within the open source Remote Sensing and GIS software Library (RSGISLib).

Keywords: image segmentation; remote sensing; land cover; iterative elimination; RSGISLib

---

1. Introduction

The classification and analysis of remotely-sensed optical data has become a key technology for the mapping and monitoring of land cover [1,2]. Traditionally, such analysis has been performed on a per pixel basis, but over the last 20 years, there has been a significant movement to embrace context and segment-based classifications [3] due to observed improvements in classification accuracy [3–7]. A significant driver for this adoption has been the availability of the eCognition software [8], which has provided user-friendly tools for these operations and made them widely available to the community. For national mapping programs, high-resolution remotely-sensed data such as RapidEye, Sentinel 1 and 2, SPOT (4–5) and Landsat (TM, ETM+, OLI), provide the majority of data. These data typically have a resolution of 5–30 m, thereby providing sufficient detail for monitoring land cover, such as forestry, agriculture and grasslands. The use of segmentation for land cover mapping at this resolution aims to identify spatially-homogeneous units such as fields and forestry blocks as single objects, or a small number of large objects. However, any under-segmentation will result in a poor classification result, as features are merged and cannot be identified. Therefore, where errors occur within the segmentation result, a small amount of over-segmentation is preferable to retain classification accuracy [9].

Building on the extensive literature on image segmentation (e.g., [10–14]) within the image processing and medical imaging communities, there have been a significant number of publications devoted to segmentation of remotely-sensed imagery. These can be categorised as either top-down or bottom-up approaches. Top-down approaches, such as multi-scale wavelet decomposition [15], start with a single region and progressively break the image into smaller regions until a termination condition is met. Bottom-up methods, such as region growing and spectral clustering [16], start with individual pixels and group them until a termination condition is met. Termination criteria vary between approaches, but they are generally a combination of the colour similarity, shape, and size of the regions generated. To group pixels within a bottom-up approach, a number of methods have been proposed, but include region growing [17], thresholding [18], statistical clustering [19], Markov random field models [20], fuzzy clustering [21], neural networks [22] and image morphology watersheds [23]. Thresholding techniques are generally applied to extract features from a distinctive background, with histograms commonly being used to identify the threshold values automatically [24]. Region growing techniques require seeds, which are difficult to generate reliably automatically and are time consuming to provide manually. Soille [17], Brunner and Soille [25] used an automated seeding method based on quasi-flat zones, as did Weber et al. [26], who applied it to multi-temporal imagery. Watershed techniques are one of the most commonly applied, but these are often over-segmented and, due to the filtering used to generate the gradient image(s) boundaries between features, are often blurred and poorly defined at the pixel level.

Segmentation techniques are often designed for specific applications, such as tree crown delineation [27], requiring considerable parameterisation that can be difficult to define and optimise [28]. The termination criteria used are also a common differentiating factor in the results, which are generated by the segmentation algorithm. For example, Baatz and Schäpe [28] aimed to minimise the spectral homogeneity of the image regions while creating regions of similar size and compact shape [8] to allow accurate statistical comparison between features for later classification. The key parameter within the eCognition multi-scale segmentation algorithm is the scale factor f [28]. The user-defined threshold of f controls the termination for the segmentation process where f is defined as:

$$
\begin{aligned}
f &= w \cdot h_{colour} + (1-w) \cdot h_{shape} \\
h_{colour} &= \sum_{b=1}^{n_{bands}} w_b \cdot \sigma_b \\
h_{shape} &= w_s \cdot h_{cmpct} + (1-w_s) \cdot h_{smooth} \\
h_{cmpct} &= \frac{p_l}{\sqrt{n_{pxl}}} \\
h_{smooth} &= \frac{p_l}{p_{bbox}}
\end{aligned}
\tag{1}
$$

where w is a user-defined weight (range: 0–1) defining the importance of object colour versus shape within the termination criteria, n_{bands} is the number of input image bands, w_b is the weight for the image band b, σ_b is the standard deviation of the object for band b, w_s is the user-defined weight (range: 0–1) defining the importance of compactness versus smoothness, p_l is the perimeter length of the object, n_{pxl} is the number of pixels within the object and p_{bbox} is the perimeter of the bounding box of the object.

Others have used k-means clustering as an alternative approach. Lucchese and Mitra [29] used k-means clustering and median filtering for post-processing. While the method is simple and effective, some of the objects with sharper edges were over-smoothed from the median filtering. Wang et al. [19] first initialised the segmentation using a k-means clustering to remove noise and reduce the feature space from single pixels to clumps. The clumps were then merged using a combination of spectral colour (weighted by object size) and object shape. The resulting segmentation is similar to that derived from the eCognition software. Likewise, Brunner and Soille [25] executed a single merge of clumps based on spectral similarity.

Computational constraints are another issue commonly encountered with many segmentation algorithms. The size of the scene that can be processed is in some cases significantly limited. Often, one whole satellite image cannot be processed in a single step, due to memory constraints. Additionally, methods commonly deployed within the field of computer vision expect either one (greyscale) or three (red, green and blue) image bands, or spectral measurements, and it is assumed that these data channels are scaled to the same range (i.e., 0–255). In remote sensing, this is commonly not the case, with many satellites having multi-spectral capabilities with at least four spectral wavelengths, including the near-infrared (NIR), which over vegetation, results in considerably larger range of values than other channels due to leaf cell structure [30].

Assessing the quality and effectiveness of a segmentation result has proven difficult [31]. In this context, the aim of segmentation is to subdivide the landscape into units of the same land cover or spectral colour, where ideally, the entire feature, such as a single species forest block, will be captured as a single segment so that these units can be classified within an appropriate context. However, if the scene is under-segmented, that is, multiple features of interest are contained within a single segment, then classification of these features is not possible. Therefore, under-segmentation is to be avoided. Therefore, a degree of over-segmentation is generally accepted, as the following classification of neighbouring objects of the same class can be merged. Additionally, the scale of the imagery and result being sought will significantly impact segmentation requirements. Segmentation quality can be assessed for specific tasks, such as delineation of tree crowns or buildings, with respect to a set of reference data. The generation of a reference set is a clear task with a relatively clear answer for specific features, but for more general segmentation tasks, such as splitting a whole SPOT 5 scene into units for land cover classification, where a large number of land covers, uses and vegetation conditions are present, there is no one "best" solution. In an attempt to quantify the differences between different segmentation results, various approaches exist [32] where parameters such as shape (e.g., shape index [33]), comparison to a set of reference features [34–36], synthetic images [37] and colour metrics [31,38] have been used.

Our aim was to provide a segmentation technique that is scalable to support the national land cover mapping program of New Zealand, while only using a few understandable parameters. Segments needed to be of uniform spectral response (i.e., colour) with a minimum size and suitable for land cover and land cover change classification, while boundaries between land covers should be sharp and accurate. The algorithm needs to be efficient over large datasets (i.e., multiple mosaicked scenes) and applicable to a wide range of sensors and image pixel resolution. We have therefore proposed a new iterative elimination algorithm to meet these aims. We describe and apply the method on Landsat ETM+ and SPOT5 imagery of typical landscapes in New Zealand.

2. Methods

The iterative elimination method we present here has four steps (Figure 1). First, is a seeding step, which identifies the unique spectral signatures within the image, second a clumping process to create unique regions, third an elimination process to remove regions below the minimum mapping unit, and, finally, the regions are relabelled so that they are sequentially numbered, providing the final segmentation. These steps require two parameters, the initial number of cluster centres to be produced and the minimum mapping unit defined for the elimination process.

When processing imagery with a large number of highly-correlated image bands (i.e., time series and hyper-spectral imagery), it is recommended that a subset or derived compression of the image data (e.g., principle component analysis) of the available input image bands be used to both minimise auto-correlation between the bands and reduce processing time. In this study, the segmentation of SPOT-5 data from two dates was undertaken using the near-infrared (NIR), shortwave infrared (SWIR) and red wavelengths (i.e., the green band was not included due to the significant correlation with the red band). Images were processed from DN to standardised reflectance before segmentation to reduce scene complexity [39].

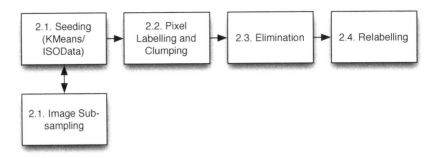

Figure 1. Flowchart for the segmentation algorithm.

2.1. Seeding

The first step in the algorithm is to seed an initial set of unique spectral signatures to represent the image. To seed, in an unsupervised fashion, the k-means [40] clustering algorithm was used. While there are many clustering algorithms within the literature such as Iterative Self-Organising Data (ISOData [41]), mean-shift [42], hierarchical clustering [43] and fuzzy k-means [44], it was found that k-means produced the best results while being computationally efficient. The k-means clustering algorithm is an iterative procedure requiring calculations in the order of nc ($\mathcal{O}(nc)$) where n is the number of pixels and c is the number of cluster centres. To minimise the number of calculations, the image was subsampled before clustering. It was found that subsampling to 10% or 1% of the image pixels was sufficient to provide similar cluster centres to those derived from the whole input dataset (Figure 2), while providing a significant performance improvement in computation time (Table 1). Another advantage of k-means is that it allows development of the model on one dataset (i.e., a subset) and application to another. As with any statistical sampling, a sufficiently large sample is required to ensure representativeness and that no features are missed. Experiments where carried out using a number of scenes (5 Landsat-7, 5 SPOT-5 and 3 WorldView-2) from each of the sensors to cover a range of environments.

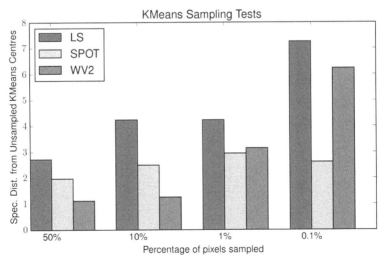

Figure 2. Demonstration that k-means cluster centres generated using 50%, 10%, 1% and 0.1% subsampled image data result in cluster centres close to the original data with considerably less computing time. The imagery used is Landsat ETM+, SPOT-5 and WorldView-2.

Table 1. Timing experiment generating 60 clusters using k-means employing resampling. Experiments were carried out on a 2.93-GHz Intel Xeon processor with 8 GB of RAM running Mac OSX 10.7 using a SPOT 5 scene with 60 million pixels and 4 images bands.

Sampling (Pixels)	Time (Minutes:Seconds)
100%	17:14
50%	11:10
20%	04:03
1%	02:39
0.1%	00:08

Within the k-means clustering process, the Euclidean distance metric is used to calculate the distance between features and cluster centres within feature space. Therefore, due to the differences in dynamic range between the input wavelengths (i.e., near-infrared has a larger range than red), some channels could be weighted higher than others and need to be rescaled. A number of options exist for rescaling the data, but it was found that normalising each image band (wavelength) independently within a range of two standard deviations from the mean, thresholded by the image minimum and maximum, produced the best results, increasing the contrast between the land cover classes.

2.2. Clumping

The k-means-classified pixels are then clumped to define a set of geographically uniquely-labelled regions. This result is over-segmented in many regions of the scene due to the per-pixel nature of the processing and spectral confusion between units. It is common that over half of the resulting clumps are only a single pixel in size (Figure 3).

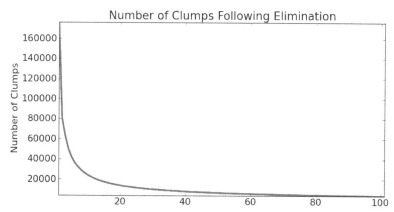

Figure 3. The number of clumps of each size at each elimination step from 1–100 pixels (1 ha) on a 10 × 10 km SPOT5 image (see Figure 4).

2.3. Local Elimination

The next stage eliminates these small clumps up to a minimum size, which will correspond with the minimum mapping unit for subsequently derived mapping products. The elimination is an iterative process (Algorithm 1) where regions are eliminated based on size, starting with the smallest (i.e., 1 pixel). To eliminate, a region is merged with its spectrally-closest (colour) neighbour, which must be larger than itself, but does not need to be larger than the minimum mapping unit. If there are no neighbouring regions larger than the region being eliminated, then it is left for the next iteration, thereby ensuring that regions are eliminated naturally into the most appropriate neighbour. A critical consideration of the elimination is that the elimination itself (i.e., merging of clumps) is not performed until the end of each iteration. Otherwise, the elimination will result in region growing from the first clump reviewed for elimination as the clump spectral mean will change. Upon merging, the clump statistics are updated to provide the new mean spectral values used within the next iteration.

Algorithm 1 Pseudocode for the elimination process.

List elimPairs
for *size* = 1 → *minunitsize* **do**
   clumps ← `GetUnitsOfSizeLessThan`(*size*)
   for all *clump* ← *clumps* **do**
      neighbors ← `GetNeighborsAreaGtThan`(*clump*, *size*)
      if *neighbors.size*() > 0 **then**
         neighbor ← `FindSpectrallyClosestNeighbor`(*neighbors*, *clump*)
         elimPairs ← *pair*(*neighbors*, *clump*)
      end if
   end for
   for all *pair* ← *elimPairs* **do**
      pair.neighbor ← `MergeRegions`(*pair.neighbor*, *pair.clump*)
   end for
end for

One problem with the elimination process is that features, which are highly spectrally distinct from their neighbours, such as water bodies, with an area smaller than the minimum clump size will be merged. This can result in undesirable spectral mixing and potentially reduces the likelihood of correct classification of land cover. To resolve this issue, an optional threshold was introduced that prevents clumps with a spectral distance (*d*) above the defined threshold from being merged with their neighbour, hence allowing regions below the minimum size to be retained. The spectral distance threshold was set at 0.1 reflectance. Whenever clumps differ by more than that, they do not merge. We found the 0.1 threshold effective, but this might need to be reviewed depending on the types of features found in some scenes.

As shown in Figure 3, the elimination process reduces the number of clumps representing a 10 × 10 km SPOT5 scene from over 170,000 to less than 4000. Before elimination (Figure 4a), there were 1137 segments over the 100-pixel (1 ha) threshold of the object size. Therefore, 31% of segments in the final result (Figure 4b) were based on a clump directly originating from the k-means clustering with the rest amalgamated from the smaller clumps, which will not correspond directly with the spectral regions defined by the k-means clustering.

The first step, which removes single pixels, results in a significant removal of clumps (170,000–80,000). For single pixels, a memory-efficient approximation of the elimination process can be adopted. Using a 3 × 3 window, all single pixel clumps can be identified and stored as a binary image mask. The single pixels can be iteratively eliminated by merging each pixel with the clump of the spectrally-closest pixel that is part of a larger clump (i.e., at least two pixels). If there is no neighbouring clump of at least two pixels, then the single pixel is considered in the next iteration. The spectrally-closest pixel may occasionally differ from the spectrally-closest clump, but this has not been found to be significant and greatly reduces the memory requirements of the elimination process, thus allowing larger datasets to be rapidly processed.

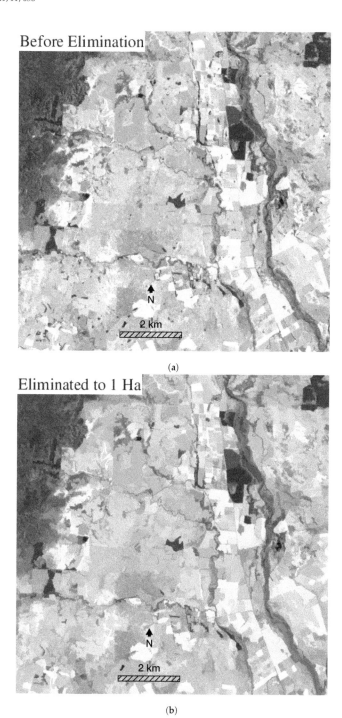

(a)

(b)

Figure 4. (**a**,**b**) Elimination from 1–100 pixels on a 10×10 km SPOT5 image (10-m resolution). Segments are coloured with their mean spectra.

2.4. Relabelling

During elimination, the features below the minimum mapping unit are removed and merged with their spectrally-closest larger neighbour. This results in a gap within the attribute table for each eliminated unit. Therefore, a relabelling process to collapse the attribute table is undertaken to ensure the feature identification numbers are sequential. Sequential numbering minimises the number of required rows within the attribute table and makes later classification more efficient.

2.5. Parameterisation

Using k-means seeding, the algorithm has two main parameters: (1) the number of seeds (k) (i.e., initial clusters in feature space) and (2) the minimum segment size for the elimination. There are no hard and fast rules for identifying these parameters, as with all segmentation algorithms, it is difficult to define a metric that completely quantifies the quality of the segmentation [31]. Nevertheless, the selection process may be guided by observing the resulting number of segments, their size, and their internal and inter-neighbour spectral variation [38]. It is worth noting that smaller segments will provide lower spectral variation both within the segments and to their neighbours.

The number of clusters selected for the k-means to be produced as seeds is the key parameter governing the spectral separation of the features. Too few clusters and features, which are close to one another within the feature space, will be under-segmented, while too many clusters will result in an over-segmented scene. Through experimentation, it was found that between 30 and 90 seeds, depending on the level of detail required by the user, were generally sufficient for multi-spectral imagery, such as RapidEye, Sentinel-2, SPOT5, WorldView-2 and Landsat (TM, ETM+, OLI), regardless of the number of image bands or the use of multi-date image stacks. Although, in some use cases, such as segmenting bright objects on a dark background (e.g., ships on the ocean), a significantly smaller number of seeds (e.g., 2 or 3) may be more suitable. For vegetation studies, a value of 60 seeds has been consistently used across a range of studies (e.g., [45,46]) using Sentinel-2, SPOT5 and Landsat data, either at a single or with multiple dates and found to produce good results.

The minimum segment size for the elimination can be related to the minimum mapping unit of the output product being derived using the resulting segments. Where a minimum mapping unit is not defined by the project, the user needs to assess the size of the smallest features they are interested in within the image. For example, if isolated trees within paddocks are of interest within high resolution imagery (i.e., WorldView2), then the minimum segment size needs to be set such that the isolated trees are above this threshold, otherwise they will be eliminated in the larger paddock segment.

3. Results

Assessment of segmentations was aided through the establishment of a set of reference regions, which were manually drawn with reference to the original satellite imagery. We also observed key features and the overall appearance of the segmentation result compared with input image. In this context, segmentation can be thought of as a means of image compression. Therefore, when segments are coloured with their mean spectral values, the resulting image should retain the same structures as the original image (i.e., as shown in Figure 4). We segmented a range of optical imagery, and the resulting number of segments, their size and their spectral variation were assessed to ascertain whether the segmentation results were fit for the purpose of land cover classification, including change.

3.1. Visual Assessment

Segmentations were produced using SPOT-5 scenes for the whole of New Zealand, and Figure 5 shows an example. The NIR, SWIR and red image bands were used and segmented with $k = 60$ and a minimum segment size of 100 pixels.

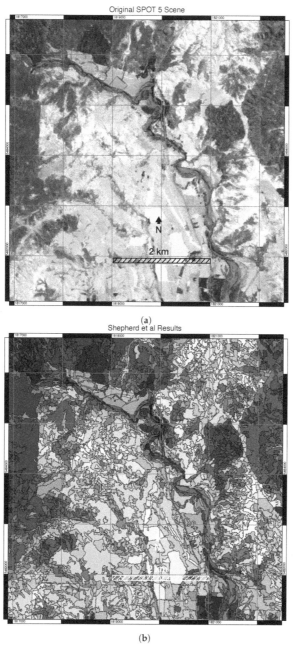

Figure 5. Example segmentation on a SPOT 5 scene (**a**), using a colour composite of red = NIR, green = SWIR and blue = Red. Segmented with 60 seeds and eliminated to a minimum mapping unit of 1 ha (**b**). The spatial pattern of land cover is well represented with the number of clumps being 0.6% of the number of pixels.

Figure 5 demonstrates the application of the segmentation process on a region of New Zealand typical to the North Island with forest, scrub, agricultural fields and grasslands at various conditions.

As demonstrated, the algorithm has maintained the spectral homogeneous forest and scrub blocks as larger continuous segments, while regions of higher spectral heterogeneity, such as the grasslands, have been segmented into smaller units. Overall, it is considered that this segmentation provides an ideal basis for a subsequent land cover classification. There is some moderate over-segmentation, particularly within the grasslands and some spectrally-heterogeneous forest, but this is often desirable to ensure under-segmentation is not present, which prevents the correct classification of those segments.

Where small spectrally-distinct features need to be retained, the spectral distance threshold on the elimination, which prevents features too far away from one another in the feature space from being merged, can be used. As demonstrated in Figure 6c, where a threshold of 0.1 reflectance has been applied, the small ponds and some additional shelter belts have been retained that are below the minimum mapping unit of 2.5 ha. This produces a segmentation result that more closely corresponds to the original image (Figure 6a) compared to the segmentation without the application of the threshold (Figure 6b).

3.2. Parameterisation

To understand the effect of the number of seeds, a number of tests were undertaken for a range of sensors. The datasets used for these experiments (Table 2) were a pair of multi-spectral WorldView2 scenes for Wales, U.K., spectrally transformed to top-of-atmosphere reflectance. A pair of SPOT5 scenes and Landsat (L4/L7), covering regions of North Island, New Zealand, scenes were all corrected to surface reflectance and standardised for Sun angle and topography [39]. For each of these sensors, experiments were undertaken for an image at a single date and the combined multi-date pair. For the single date tests, all the image bands were used, while for the multi-date tests, a subset of the image bands deemed optimal for visualisation was used. For the SPOT5 and Landsat data, these were the red, NIR and SWIR (SWIR1 for Landsat) image bands; while for WorldView2 images, bands NIR (Band 8) , red edge (Band 6) and blue (Band 2) were used. The minimum segment size was defined as 100 pixels for the WorldView2 and SPOT5 data and 30 pixels for Landsat 4/7 data.

Original SPOT 5

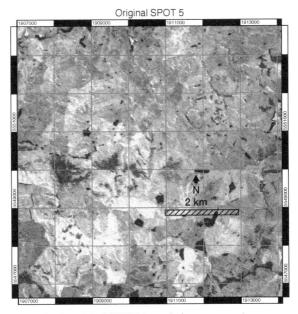

(**a**) The original SPOT-5 image before segmentation.

Figure 6. *Cont.*

Segmentation on SPOT 5 Data with No Spectral Threshold

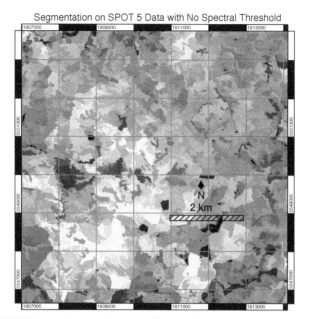

(**b**) The segmented SPOT-5 image, shown as the mean-reflectance for a segment, without using a spectral distance threshold to allow smaller features to be retained.

Segmentation on SPOT 5 Data with Spectral Threshold of 0.1

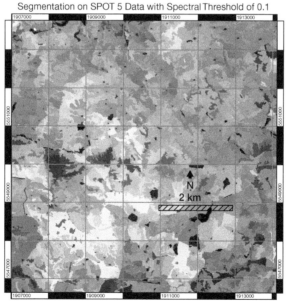

(**c**) The segmented SPOT-5 image, shown as the mean-reflectance for a segment, where a threshold to allow smaller features to be retained has been used.

Figure 6. (**a**) The original SPOT 5 image, (**b**) the segmentation applied without a spectral distance threshold (60 seeds and eliminated to the 2.5-ha minimum mapping unit) and (**c**) the segmentation applied with a spectral distance threshold of 0.1 reflectance. Images are displayed with a band combination of NIR, SWIR and red.

Table 2. Datasets used for parameter testing. * indicates the image used for the single date tests.

Sensor	Year (s)	Image Size	Pixel Resolution
Worldview2	July 2011 *, November 2011	3539×3660	2 m
SPOT5	2008, 2012 *	7753×7703	10 m
Landsat 4/7	1990, 2002 *	$12,000 \times 12,000$	15 m

Figure 7 shows the number of segments required to characterise 50% of the total area of the scenes. If a large number of segments are required to make up 50% of the area, it is likely the scene is over-segmented, while if small, it is likely to be under-segmented. These plots therefore provide an understanding of the relationship between the number and size of the segments with respect to the number of k clusters, from which the recommendation that the suitable range of values for k is 30–90 can be made. A concurrent visual examination of segments is important.

For the SPOT scenes (Figure 7A,B), the number of segments increased rapidly before plateauing. This shape suggests there is limited benefit in increasing the size of k once the plateau has been reached. In the test datasets, it was observed that a typical range of k was from 30–90 clusters (shown by the red lines). The 50% cumulative area profile (Figure 7C) for the WorldView2 scene (Figure 8) is a straight line. This is due to the number of large features within the scene, specifically the raised bog (Borth Bog, Wales, U.K.) in the centre of the scene, the estuary and sea, which even at high values of k, generated large segments due to their spectral homogeneity. When higher values of the cumulative area were considered (Figure 7D), the normal profile was seen, and the range of k from 30–90 clusters when visually assessed was still found to be appropriate.

3.3. Comparison to Other Algorithms

To compare the performance of the algorithm detailed in this paper to those of others, four alternatives were identified. The algorithms used for the comparison were the mean-shift algorithm [42] implemented within the Orfeo toolbox [47], as it is widely cited (e.g., [48–50]) as an approach that produced good results on a wide variety of Earth Observation (EO) data, the eCognition multi-resolution segmentation algorithm [28], as the algorithm most commonly used within the literature (e.g., [1,2,51]), and the Quickshift algorithm of Vedaldi and Soatto [52] and the algorithm of Felzenszwalb and Huttenlocher [53], implemented within the scikit-image library and interfaced within the RSGISLib library [54], as examples of more recent approaches from the computer vision community applicable to EO data. A SPOT-5 scene (a subset of which is shown in Figure 10A), which represents a range of land covers and uses, was used for the experiment. To identify the parameters for each of the segmentation algorithms, a grid search was performed for each algorithm across a range of parameters (Table 3).

Table 3. Parameters used for each of the segmentation algorithms.

Algorithm	Parameters	Number of Segmentations
eCognition	scale: [10–100], shape: [0–1], compact.: [0–1]	1210
Mean-Shift	range radius; [5–25], convergence thres.: [0.01–0.5], max. iter.: [10–500], min. size: [10–500]	625
Felzenszwalb	scale: [0.25–10], sigma: [0.2–1.4], min. size: [5–500]	343
Quickshift	ratio: [0.1–1], kernel size: [1–20], max. dist.: [1–30], sigma: [0–5]	1500
Shepherd et al.	k: [5–120], d: [10–10,000], min. size: [5–200]	540

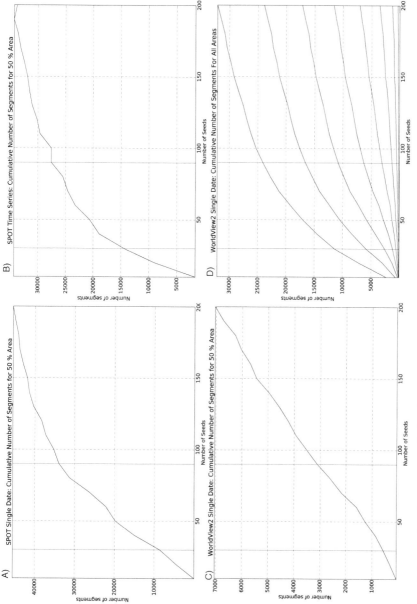

Figure 7. Cumulative number of segments required to characterise 50% of the area of (**A**) the single date SPOT5 scene, (**B**) multi-date SPOT5 time series, (**C**) the single date WorldView2 image and (**D**) the number of segments required to characterise 10–90% of the area of a single–date WorldView2 image (at 10% increments; the top line is 90%).

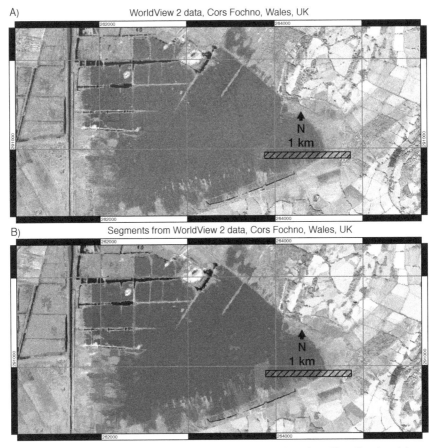

Figure 8. (**A,B**) Example segmentation on a WorldView2 scene, with a value of $k = 60$ and a minimum mapping unit of 0.4 ha, using a colour composite of red = NIR, green = red edge and blue = blue. The spatial pattern of land cover is well represented with the number of clumps being 0.2% of the number of pixels.

To compare the parameterisations and segmentation quality, two quantitative approaches were adopted. The first used a set of 200 reference segments (Figure 9), manually drawn on the SPOT-5 image, to calculate the precision and recall metrics of Zhang et al. [36]. The precision and recall metrics were combined to produce the f metric [36] as follows using an α of 0.5:

$$f = \frac{1}{\alpha \frac{1}{\text{precision}} + (1 - \alpha) \frac{1}{\text{recall}}} \quad (2)$$

The second approach was to use the metrics of Johnson and Xie [38] for estimating the over- and under-segmentation within the result. For this study, the metrics of Johnson and Xie [38] were calculated with all four spectral bands of SPOT-5 to create the overall global score (gs). To combine f and gs to generate a single ranked list, gs was normalised to a range of 0–1 (gs_norm) and added to f, using a weight ω:

$$gs\_f = \omega gs\_norm + (1 - \omega)f \quad (3)$$

For this analysis, ω was set at 0.25, giving more weight to reference segments metric f. The top parameter sets and metrics for each segmentation algorithm are given in Table 4.

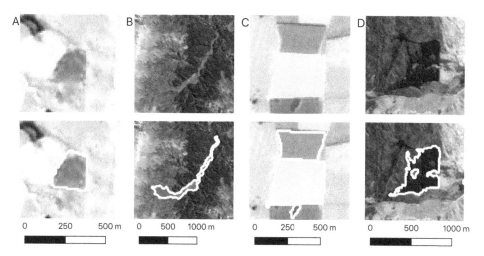

Figure 9. (A–D) Examples of the reference segments that were manually drawn.

Table 4. Comparison of the segmentation results.

Algorithm	Parameters	Rank	gs_f	f	Precision	Recall	gs	gs_{green}	gs_{red}	gs_{nir}	gs_{swir}
Shepherd et al.	k: 60, d: 10,000, min. size: 10	1	0.74	0.85	0.84	0.86	0.97	1.01	0.90	0.99	0.98
Quickshift	ratio: 0.75, kernel size: 10, max. dist.: 5, sigma: 0, lab colour space.	86	0.64	0.73	0.80	0.68	0.94	0.92	0.86	0.99	0.98
Mean-Shift	range radius; 15, convergence thres.: 0.2, max. iter.: 100, min. size: 10	253	0.56	0.61	0.55	0.70	0.93	0.95	0.87	0.96	0.94
eCognition	scale: 10, shape: 0.7, compact.: 0.2	411	0.49	0.52	0.57	0.48	0.95	0.99	0.92	0.95	0.93
Felzenszwalb	scale: 10, sigma: 12, min. size: 20	539	0.47	0.46	0.49	0.43	1.10	1.14	1.04	1.17	1.04

The rank refers to the position in the order list of all performed segmentations. Therefore, using the gs_f metric, there were 85 segmentations using different parameterisations of the Shepherd et al. algorithm (from this article) that were ranked higher than the highest Quickshift segmentation. For the mean-shift, there were 252 Quickshift and Shepherd et al. segmentations ranked higher than the highest mean-shift. Of those Shepherd et al. segmentations, the parameters, particularly k, were clustered with a k value of 60 being identified for the top 38 segmentations, corresponding with the analysis shown in Figure 7.

Figure 10 illustrates the segmentation results from the comparison using the parameters outlined in Table 4. The results were similar; however, two were visually more appropriate, these being the Shepherd et al. (Figure 10B) and mean-shift (Figure 10D), as they both captured the majority of the homogeneous regions as continuous segments while sub-dividing the areas of spectral variation. Both the Shepherd et al. (Figure 10B) and mean-shift (Figure 10D) algorithms demonstrated some artefacts associated with the smooth gradient transitions, as do the other methods tested. In regions of transition, the Quickshift algorithm (Figure 10C) produced a large number of very small segments, and these are not desirable within a segmentation approach; however, the homogeneous regions were well delineated with good correspondence with the reference regions. The eCognition multi-resolution segmentation (Figure 10E) aims to produce segments of similar size, and therefore, had some over-segmentation of the homogeneous features present in the Shepherd et al. (Figure 10B), Quickshift (Figure 10C) and mean-shift (Figure 10D) segmentations.

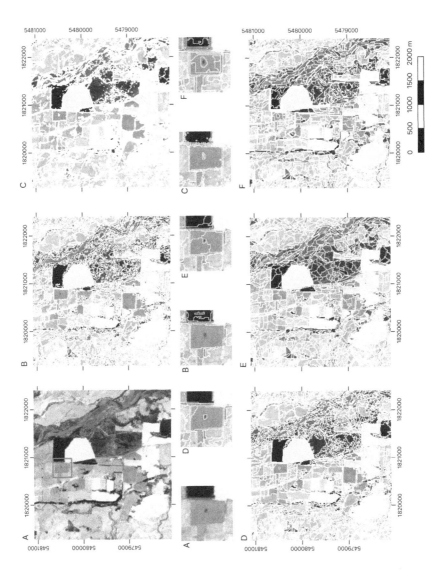

Figure 10. Comparison of segmentation algorithms with the parameters found in Table 4; the red box in (**A**) highlights the region of the zoom-in area. (**A**) SPOT-5 image with NIR, SWIR, red band combination, (**B**) Shepherd et al. (this article), (**C**) Quickshift [52], (**D**) mean-shift [42], (**E**) eCognition [28] and (**F**) Felzenszwalb [53].

4. Discussion

It is difficult to compare segmentation algorithms because comparison involves multiple criteria and depends on the specific application and scale [51]. However, we have demonstrated that the algorithm proposed in this paper compares favourably in most assessment metrics with the commonly-used mean-shift [42] and multi-resolution [28] segmentation approaches when applied to a typical rural and forest landscape in New Zealand. In addition, our algorithm has attributes that make it more useful in certain applications. It is readily scalable to large areas, such as nations or regions (e.g., [55]), which is desirable for preventing hard boundaries on tiles. It uses a small number of parameters, which may be consistently used across a large range of geographic areas and data types. It permits the direct setting of a minimum mapping unit, and it is freely available in open-source software.

The technique presented in this paper is being used operationally for national land-cover and land-cover change mapping of New Zealand using a single parametrisation (60 seeds and 100-pixel minimum mapping unit) on SPOT-5 and Sentinel-2 data. This is because the iterative elimination algorithm is highly scalable, being operationally applied to each of the regions of New Zealand, the largest of which is Canterbury. The Canterbury regional mosaic comprised 36 SPOT5 scenes and produced a single image with 36,533 × 35,648 pixels. Computationally, the segmentation of the Canterbury region required 12 GB of RAM and ran in approximately three hours on a 3-GHz x86 processor running Linux, producing 1,222,885 segments. This level of scalability is advantageous for operational use, as image tiling is not required, thus avoiding complex and time-consuming operations to merge tile boundaries, and the results are consistent across the scene. However, for larger areas, we have also implemented a tiled version of the algorithm [56], which has been used to produce a single segmentation for the Australian continent [55] using a mosaic (141,482 × 130,103 pixels) of Landsat and ALOS PALSAR. This resulted in 33.7 million segments.

In hilly and mountainous terrain topographic shadowing can significantly affect the result of the image segmentation, as the shadowing alters the spectral response of those pixels. We therefore recommend that imagery be standardised to a consistent Sun and view angle before segmentation [39]. For example, a SPOT5 scene of 7753 × 7703 pixels, segmented with 60 clusters and eliminated to 100 pixels, generated 7715 fewer segments when standardised imagery was used, but more significantly, the distribution of those features only corresponded with patterns in land cover as opposed to land cover and topographic shadowing.

We have also successfully applied our segmentation algorithm to multi-date imagery. In this case, the segmentation is required to capture the information present in both scenes and the change between the scenes for subsequent classification. The change segments can be easily differentiated from the rest by the change in spectral reflectance between early and later dates. Automatic identification of these change segments has proven useful in developing a semi-automated method for updating land cover.

Segments are now being successfully provided across large regions using this algorithm. Therefore, new methods of assessing the quality of segmentation results are required, so that optimal parameters can be estimated. This will become increasingly important as datasets for segmentation become larger. Additionally, when segmenting large regions, the number of segments being generated is large, and their classification remains challenging, primarily due to the significant memory requirements for attribution. Existing software and methods are commonly unable to cope with this number of segments. Therefore, a new and advanced attribute table file and image format (using HDF5, called KEA [57]) and software API have been implemented within the RSGISLib [54] (http://www.rsgislib.org) to support rule-based object-oriented classification of the segments. For full details, see Clewley et al. [56].

5. Conclusions

This paper has outlined a new technique for the segmentation of remotely-sensed imagery, which is highly scalable through an innovative iterative elimination process. It is being used

Remote Sens. **2019**, *11*, 658

operationally at national scales across New Zealand. The technique has clear and simple parameters and can be consistently applied across a large range of geographic areas and data types. Segmentation quality is similar to that achieved by other commonly-used algorithms such as mean-shift implemented within the Orfeo toolbox and the multi-resolution segmentation algorithm widely used within the eCognition software. The algorithm provides good performance in both memory usage and computation time. The algorithm is recommended for segmenting all modalities of remotely-sensed imagery, although pre-processing of SAR data for speckle reduction is required. Additionally, a free and open implementation of the algorithm has also been provided as part of the remote sensing and GIS software library (RSGISLib; [54]) and can be downloaded from http://www.rsgislib.org.

Author Contributions: J.D.S. was responsible for algorithm development. P.B. was responsible for algorithm application. J.R.D. contributed to algorithm application. All authors were involved in reviewing and approving the final version of the manuscript.

Funding: The Ministry of Business Innovation & Employment (New Zealand) funded this research under Contract C09X1709.

Conflicts of Interest: The authors declare no conflict of interest.

References

1. Lucas, R.; Rowlands, A.; Brown, A.; Keyworth, S.; Bunting, P. Rule-based classification of multi-temporal satellite imagery for habitat and agricultural land cover mapping. *ISPRS J. Photogramm. Remote Sens.* **2007**, *62*, 165–185. [CrossRef]

2. Lucas, R.; Medcalf, K.; Brown, A.; Bunting, P.; Breyer, J.; Clewley, D.; Keyworth, S.; Blackmore, P. Updating the Phase 1 habitat map of Wales, UK, using satellite sensor data. *ISPRS J. Photogramm. Remote Sens.* **2011**, *66*, 81–102. [CrossRef]

3. Blaschke, T. Object based image analysis for remote sensing. *ISPRS J. Photogramm. Remote Sens.* **2010**, *65*, 2–16. [CrossRef]

4. Lobo, A. Image segmentation and discriminant analysis for the identification of land cover units in ecology. *IEEE Trans. Geosci. Remote Sens.* **1997**, *35*, 1136–1145. [CrossRef]

5. Stuckens, J.; Coppin, P.; Bauer, M. Integrating contextual information with per-pixel classification for improved land cover classification. *Remote Sens. Environ.* **2000**, *71*, 282–296. [CrossRef]

6. Fuller, R.; Smith, G.; Sanderson, J.; Hill, R.; Thomson, A. The UK Land Cover Map 2000: Construction of a parcel-based vector map from satellite images. *Cartogr. J.* **2002**, *39*, 15–25. [CrossRef]

7. Tarabalka, Y.; Benediktsson, J.; Chanussot, J. Spectral-Spatial Classification of Hyperspectral Imagery Based on Partitional Clustering Techniques. *IEEE Trans. Geosci. Remote Sens.* **2009**, *47*, 2973–2987. [CrossRef]

8. Definiens. *eCognition Version 5 Object Oriented Image Analysus User Guide*; Technical Report; Definiens AG: Munich, Germany, 2005.

9. Carleer, A.P.; Debeir, O.; Wolff, E. Assessment of Very High Spatial Resolution Satellite Image Segmentations. *Photogramm. Eng. Remote Sens.* **2005**, *71*. [CrossRef]

10. Fu, K.S.; Mui, J.K. A survey on image segmentation. *Pattern Recognit.* **1981**, *13*, 3–16. [CrossRef]

11. Haralick, R.M.; Shapiro, L.G. Image segmentation techniques. *Comput. Vis. Graph. Image Process.* **1985**, *29*, 100–132. [CrossRef]

12. Pal, N.; Pal, S. A review on image segmentation techniques. *Pattern Recognit.* **1993**, *26*, 1277–1294. [CrossRef]

13. Cheng, H.; Jiang, X.; Sun, Y.; Wang, J. Color image segmentation: Advances and prospects. *Pattern Recognit.* **2001**, *34*, 2259–2281. [CrossRef]

14. Dey, V.; Zhang, Y.; Zhong, M. A review on image segmentation techniques with remote sensing perspective. In Proceedings of the ISPRS TC VII Symposium—100 Years ISPRS, Vienna, Austria, 5–7 July 2010; Volume 38, pp. 5–7.

15. Chen, Q.; Luo, J.; Zhou, C.; Pei, T. A hybrid multi-scale segmentation approach for remotely sensed imagery. In Proceedings of the 2003 IEEE International Geoscience and Remote Sensing Symposium, Toulouse, France, 21–25 July 2003; Volume 6, pp. 3416–3419.

16. Tilton, J.C. Image segmentation by region growing and spectral clustering with a natural convergence criterion. In Proceedings of the IEEE International Geoscience and Remote Sensing Symposium Proceedings (IGARSS), Seattle, WA, USA, 6–10 July 1998; Volume 4, pp. 1766–1768.

17. Soille, P. Constrained connectivity for hierarchical image partitioning and simplification. *IEEE Trans. Pattern Anal. Mach. Intell.* **2008**, *30*, 1132–1145. [CrossRef] [PubMed]

18. Ghamisi, P. A Novel Method for Segmentation of Remote Sensing Images based on Hybrid GAPSO. *Int. J. Comput. Appl.* **2011**, *29*, 7–14. [CrossRef]

19. Wang, Z.; Jensen, J.R.; Im, J. An automatic region-based image segmentation algorithm for remote sensing applications. *Environ. Model. Softw.* **2010**, *25*, 1149–1165. [CrossRef]

20. Yang, Y.; Han, C.; Han, D. A Markov Random Field Model-based Fusion Approach to Segmentation of SAR and Optical Images. In Proceedings of the IEEE International Geoscience and Remote Sensing Symposium (IGARSS), Boston, MA, USA, 7–11 July 2008; Volume 4.

21. Zhao, F.; Jiao, L.; Liu, H. Spectral clustering with fuzzy similarity measure. *Digit. Signal Process.* **2011**, *21*, 701–709. [CrossRef]

22. Tasdemir, K. Neural network based approximate spectral clustering for remote sensing images. In Proceedings of the IEEE International Geoscience and Remote Sensing Symposium (IGARSS), Vancouver, BC, Canada, 24–29 July 2011; pp. 2884–2887.

23. Tarabalka, Y.; Chanussot, J.; Benediktsson, J. Segmentation and classification of hyperspectral images using watershed transformation. *Pattern Recognit.* **2010**, *43*, 2367–2379. [CrossRef]

24. El Zaart, A.; Ziou, D.; Wang, S.; Jiang, Q. Segmentation of SAR images. *Pattern Recognit.* **2002**, *35*, 713–724. [CrossRef]

25. Brunner, D.; Soille, P. Iterative area filtering of multichannel images. *Image Vis. Comput.* **2007**, *25*, 1352–1364. [CrossRef]

26. Weber, J.; Petitjean, F.; Gançarski, P. Towards Efficient Satellite Image Time Series Analysis: Combination of Dynamic Time Warping and Quasi-Flat Zones. In Proceedings of the IEEE International Geoscience and Remote Sensing Symposium (IGARSS), Munich, Germany, 22–27 July 2012.

27. Culvenor, D. TIDA: An algorithm for the delineation of tree crowns in high spatial resolution remotely sensed imagery. *Comput. Geosci.* **2002**, *28*, 33–44. [CrossRef]

28. Baatz, M.; Schäpe, A. Multiresolution segmentation: An optimization approach for high quality multi-scale image segmentation. *J. Photogramm. Remote Sens.* **2000**, *58*, 12–23.

29. Lucchese, L.; Mitra, S.K. Unsupervised segmentation of color images based on k-means clustering in the chromaticity plane. In Proceedings of the IEEE Workshop on Content-Based Access of Image and Video Libraries (CBAIVL'99), Fort Collins, CO, USA, 22 June 1999.

30. Rouse, J.; Haas, R.; Schell, J.; Deering, D. Monitoring vegetation systems in the Great Plains with ERTS. In Proceedings of the NASA Goddard Space Flight Center, 3d ERTS-1 Symposium (NASA SP-351 I), Greenbelt, MD, USA, 1 January 1974; pp. 309–317.

31. Zhang, H.; Fritts, J.E.; Goldman, S.A. Image segmentation evaluation: A survey of unsupervised methods. *Comput. Vis. Image Underst.* **2008**, *110*, 260–280. [CrossRef]

32. Costa, H.; Foody, G.M.; Boyd, D. Supervised methods of image segmentation accuracy assessment in land cover mapping. *Remote Sens. Environ.* **2018**, *205*, 338–351. [CrossRef]

33. Novack, T.; Fonseca, L.; Kux, H. Quantitative comparison of segmentation results from ikonos images sharpened by different fusion and interpolation techniques. In Proceedings of the GEOgraphic Object Based Image Analysis for the 21st Century (GEOBIA), Calgary, AB, Canada, 5–8 August 2008.

34. Neubert, M.; Herold, H. Evaluation of remote sensing image segmentation quality—Further results and concepts. In Proceedings of the Bridging Remote Sensing and GIS 1st International Conference on Object-Based Image Analysis (OBIA), Salzburg, Austria, 4–5 July 2006.

35. Neubert, M.; Herold, H. Assessment of remote sensing image segmentation quality. In Proceedings of the GEOgraphic Object Based Image Analysis for the 21st Century (GEOBIA), Calgary, AB, Canada, 5–8 August 2008.

36. Zhang, X.; Feng, X.; Xiao, P.; He, G.; Zhu, L. Segmentation quality evaluation using region-based precision and recall measures for remote sensing images. *ISPRS J. Photogramm. Remote Sens.* **2015**, *102*, 73–84. [CrossRef]

37. Marcal, A.R.S.; Rodrigues, S.A. A framework for the evaluation of multi-spectral image segmentation. In Proceedings of the GEOgraphic Object Based Image Analysis for the 21st Century (GEOBIA), Calgary, AB, Canada, 5–8 August 2008.

38. Johnson, B.; Xie, Z. Unsupervised image segmentation evaluation and refinement using a multi-scale approach. *ISPRS J. Photogramm. Remote Sens.* **2011**, *66*, 473–483. [CrossRef]

39. Shepherd, J.D.; Dymond, J.R. Correcting satellite imagery for the variance of reflectance and illumination with topography. *Int. J. Remote Sens.* **2003**, *24*, 3503–3514. [CrossRef]

40. MacQueen, J. Some methods for classification and analysis of multivariate observations. In *Fifth Berkeley Symposium on Mathematical Statistics and Probability*; University of California Press: Berkeley, CA, USA, 1967; Volume 1, pp. 281–297.

41. Ball, G.; Hall, D. A clustering technique for summarizing multivariate data. *Behav. Sci.* **1967**, *12*, 153–155. [CrossRef]

42. Comaniciu, D.; Meer, P. Mean shift: A robust approach toward feature space analysis. *IEEE Trans. Pattern Anal. Mach. Intell.* **2002**, *24*, 603–619. [CrossRef]

43. Ward, J.H., Jr. Hierarchical Grouping to Optimize an Objective Function. *J. Am. Stat. Assoc.* **1963**, *58*, 236–244. [CrossRef]

44. Bezdek, J.C. Fuzzy Mathematics in Pattern Classification. Ph.D. Thesis, Cornell University, Ithaca, NY, USA, 1963.

45. Dymond, J.R.; Shepherd, J.D.; Newsome, P.F.; Gapare, N.; Burgess, D.W.; Watt, P. Remote sensing of land-use change for Kyoto Protocol reporting: The New Zealand case. *Environ. Sci. Policy* **2012**, *16*, 1–8. [CrossRef]

46. Lucas, R.; Clewley, D.; Accad, A.; Butler, D.; Armston, J.; Bowen, M.; Bunting, P.; Carreiras, J.; Dwyer, J.; Eyre, T.; et al. Mapping forest growth stage in the Brigalow Belt Bioregion of Australia through integration of ALOS PALSAR and Landsat-derived Foliage Projected Cover (FPC) data. *Remote Sens. Environ.* **2014**, *155*, 42–57. [CrossRef]

47. Michel, J.; Feuvrier, T.; Inglada, J. Reference algorithm implementations in OTB: Textbook cases. *IEEE Int. Geosci. Remote Sens. Symp.* **2009**. [CrossRef]

48. Chaabouni-Chouayakh, H.; Datcu, M. Coarse-to-fine approach for urban area interpretation using TerraSAR-X data. *IEEE Geosci. Remote Sens. Soc. Newsl.* **2010**, *7*, 78–82. [CrossRef]

49. Chehata, N.; Orny, C.; Boukir, S.; Guyon, D.; Wigneron, J.P. Object-based change detection in wind storm-damaged forest using high-resolution multispectral images. *Int. J. Remote Sens.* **2014**, *35*, 4758–4777. [CrossRef]

50. Bellakanji, A.C.; Zribi, M.; Lili-Chabaane, Z.; Mougenot, B. Forecasting of Cereal Yields in a Semi-arid Area Using the Simple Algorithm for Yield Estimation (SAFY) Agro-Meteorological Model Combined with Optical SPOT/HRV Images. *Sensors* **2018**, *18*, 2138. [CrossRef]

51. Mikes, S.; Haindl, M. Benchmarking of remote sensing segmentation methods. *IEEE J. Sel. Top. Appl. Earth Obs. Remote Sens.* **2015**, *8*, 2240–2248. [CrossRef]

52. Vedaldi, A.; Soatto, S. Quick shift and kernel methods for mode seeking. In *European Conference on Computer Vision*; Springer: Berlin/Heidelberg, Germany, 2008.

53. Felzenszwalb, P.; Huttenlocher, D. Efficient graph-based image segmentation. *Int. J. Comput. Vis.* **2004**, *59*, 167–181. [CrossRef]

54. Bunting, P.; Clewley, D.; Lucas, R.M.; Gillingham, S. The Remote Sensing and GIS Software Library (RSGISLib). *Comput. Geosci.* **2014**, *62*, 216–226. [CrossRef]

55. Scarth, P.; Armston, J.; Lucas, R.; Bunting, P. A Structural Classification of Australian Vegetation Using ICESat/GLAS, ALOS PALSAR, and Landsat Sensor Data. *Remote Sens.* **2019**, *11*, 147. [CrossRef]

56. Clewley, D.; Bunting, P.; Shepherd, J.; Gillingham, S.; Flood, N.; Dymond, J.; Lucas, R.; Armston, J.; Moghaddam, M. A Python-Based Open Source System for Geographic Object-Based Image Analysis (GEOBIA) Utilizing Raster Attribute Tables. *Remote Sens.* **2014**, *6*, 6111–6135. [CrossRef]

57. Bunting, P.; Gillingham, S. The KEA image file format. *Comput. Geosci.* **2013**, *57*, 54–58. [CrossRef]

Direct, ECOC, ND and END Frameworks—Which One Is the Best? An Empirical Study of Sentinel-2A MSIL1C Image Classification for Arid-Land Vegetation Mapping in the Ili River Delta, Kazakhstan

Alim Samat [1,2,3,*], Naoto Yokoya [4], Peijun Du [5], Sicong Liu [6], Long Ma [1,2,3], Yongxiao Ge [1,2], Gulnura Issanova [7,8], Abdula Saparov [9], Jilili Abuduwaili [1,2,3] and Cong Lin [5]

1   State Key Laboratory of Desert and Oasis Ecology, Xinjiang Institute of Ecology and Geography, CAS, Urumqi 830011, China
2   Research Center for Ecology and Environment of Central Asia, CAS, Urumqi 830011, China
3   University of Chinese Academy of Sciences, Beijing 100049, China
4   RIKEN Center for Advanced Intelligence Project, Tokyo 103-0027, Japan
5   Department of Geographical Information Science, Nanjing University, Nanjing 210093, China
6   College of Surveying and Geoinformatics, Tongji University, Shanghai 200092, China
7   Faculty of Geography and Environmental Sciences, Al-Farabi Kazakh National University, Almaty 050040, Kazakhstan
8   Research Center for Ecology and Environment of Central Asia (Almaty), CAS, Almaty 050060, Kazakhstan
9   U.U. Uspanov Kazakh Research Institute of Soil Science and Agrochemistry, Almaty 050060, Kazakhstan
*   Correspondence: alim_smt@ms.xjb.ac.cn; Tel: +86-0991-782-7371

Received: 30 July 2019; Accepted: 16 August 2019; Published: 20 August 2019

check for
updates

Abstract: To facilitate the advances in Sentinel-2A products for land cover from Moderate Resolution Imaging Spectroradiometer (MODIS) and Landsat imagery, Sentinel-2A MultiSpectral Instrument Level-1C (MSIL1C) images are investigated for large-scale vegetation mapping in an arid land environment that is located in the Ili River delta, Kazakhstan. For accurate classification purposes, multi-resolution segmentation (MRS) based extended object-guided morphological profiles (EOMPs) are proposed and then compared with conventional morphological profiles (MPs), MPs with partial reconstruction (MPPR), object-guided MPs (OMPs), OMPs with mean values (OMPsM), and object-oriented (OO)-based image classification techniques. Popular classifiers, such as C4.5, an extremely randomized decision tree (ERDT), random forest (RaF), rotation forest (RoF), classification via random forest regression (CVRFR), ExtraTrees, and radial basis function (RBF) kernel-based support vector machines (SVMs) are adopted to answer the question of whether nested dichotomies (ND) and ensembles of ND (END) are truly superior to direct and error-correcting output code (ECOC) multiclass classification frameworks. Finally, based on the results, the following conclusions are drawn: 1) the superior performance of OO-based techniques over MPs, MPPR, OMPs, and OMPsM is clear for Sentinel-2A MSIL1C image classification, while the best results are achieved by the proposed EOMPs; 2) the superior performance of ND, ND with class balancing (NDCB), ND with data balancing (NDDB), ND with random-pair selection (NDRPS), and ND with further centroid (NDFC) over direct and ECOC frameworks is not confirmed, especially in the cases of using weak classifiers for low-dimensional datasets; 3) from computationally efficient, high accuracy, redundant to data dimensionality and easy of implementations points of view, END, ENDCB, ENDDB, and ENDRPS are alternative choices to direct and ECOC frameworks; 4) surprisingly, because in the ensemble learning (EL) theorem, "weaker" classifiers (ERDT here) always have a better chance of reaching the trade-off between diversity and accuracy than "stronger" classifies (RaF, ExtraTrees, and SVM here), END with ERDT (END-ERDT) achieves the best performance with less than a 0.5% difference in the overall accuracy (OA) values, but is 100 to 10000 times faster than END with RaF and ExtraTrees, and ECOC with SVM while using different datasets with various dimensions; and, 5) Sentinel-2A

MSIL1C is better choice than the land cover products from MODIS and Landsat imagery for vegetation species mapping in an arid land environment, where the vegetation species are critically important, but sparsely distributed.

Keywords: ND; END; ECOC; MRS; Extended object-guided morphological profiles; Multiclass classification; Arid-land vegetation mapping; Sentinel-2A MSIL1C; Central Asia

1. Introduction

Arid and semiarid lands encompass approximately 30–40% of the Earth's surface, and Central Asia contains one of the world's largest arid and semiarid areas. These areas have harsh climatic conditions and they are under high pressure to produce food and fibers for their rapidly increasing populations, which include a wide range of land utilization and management regimes, which results in a reduction in arid ecosystem quality. Understanding the effects and responses between landscapes and regional environments is fundamental to maintain their ecological and productive value in such circumstances. Hence, the effects and responses of landscape heterogeneity on the local and regional atmosphere, the surface energy balance, the carbon exchange, and climate changes are major topics that have attracted widespread interest [1–5]. Among these responses, the vegetation species, distribution, diversity, and biomass in these lands typically undergo wide seasonal and international fluctuations, which are largely regulated by water availability and impacted by both climatic shifts and human activities [6–8]. Thus, monitoring the vegetation status of these lands is an essential part of identifying problems, developing solutions, and assessing the effects of actions.

However, large-scale and long-term field sampling of vegetation information is challenging when considering the sampling efforts and costs. Moreover, the samples are often very sparsely distributed, and site revisits remain infrequent, while the success of any monitoring of vegetation dynamics depends on the availability of up-to-date and spatially detailed species richness and distribution at a regional scale [9–11]. Fortunately, satellite-based remote sensing (RS) data can address these challenges by identifying and detailing the biophysical characteristics of vegetation species' habitats, predicting the distribution and spatial variability in the richness, and detecting natural and human-caused changes at scales that range from individual landscapes to the whole world [1,9,12–18]. Therefore, an increasing number of geologists, ecologists, and biologists are turning to rapidly develop RS data sources for vegetation-based ecological and environmental research at local, regional, and global scales [19–24].

Regarding applications of RS data in vegetation studies, high- and moderate-resolution optical RS sensors, including IKONOS, Satellite for Observation of Earth (SPOT), Thematic Mapper ™, Enhanced Thematic Mapper (ETM), ETM+, Operational Land Imager (OLI), Sentinel-2, and Moderate Resolution Imaging Spectroradiometer (MODIS), are widely accepted and are considered to be adequate for vegetation species, diversity, distribution, and biophysical information extraction in different settings [25–32]. Creating land cover maps that detail the biophysical cover of the Earth's surface is among the oldest and ongoing hot applications. Land cover maps have been continuously suggested proven especially valuable for predicting the distribution of both individual plant species and species assemblages across broad areas that could not otherwise be surveyed in more quantitative ways with respect to vegetation index (VI)-based approaches [9,33–35]. In particular, after various land cover products that are derived from RS data at the regional and global scales have been produced, and they are freely available at spatial resolutions from 30 m to 1 km. Solid proofs can be found for extensive applications of representative products, including the 1 km University of Maryland (UMD) land cover layer [36], the Global Land Cover 2000 (GLC2000) products [37], the MODIS products [38], the 500 m MODIS [39], the 300 m GlobCover [40], and the 30 m global land cover maps [41] for vegetation studies at the regional and global scales [42–49]. However, most of the existing land cover products are coarse, not only in the spatial resolution and land cover type details, but also in the update frequency. For example, the 30 m global land cover maps are only available for 2010, 2015, and 2017 with a maximum of 10 land cover types (only eight types for our study area), while the MODIS products are only available for 2000, 2005, 2010, and 2015 with a maximum 300 m resolution (only 15 types for our study area). Furthermore, the differences between these products are very large, as shown in Figures 1b and 1c.

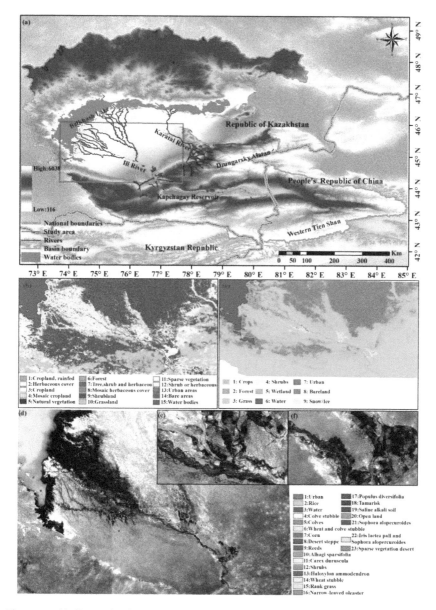

Figure 1. (**a**) Geographic location of the study area, (**b**) 2015 Moderate Resolution Imaging Spectroradiometer land use and cover change (MODIS LUCC), (**c**) 2017 GLC30, (**d**) blue rectangle, (**e**) blue rectangle, and (**f**) green rectangle. Sentinel-2 RGB image of the study area with in situ points (red dots) and the corresponding land cover types.

The Sentinel-2 mission comprises a constellation of two polar-orbiting satellites placed in the same orbit with a five-day revisit time over land and coastal areas. Each satellite carries a MultiSpectral Instrument (MSI) with 13 bands spanning from the visible and the near-infrared (VNIR) portion of the electromagnetic spectrum to the short wave infrared (SWIR) portion of the spectrum and features four bands at a 10 m spatial resolution, six bands at a 20 m spatial resolution, and three bands at a 60 m spatial resolution with a 290 km field of view [50]. Since Sentinel-2A and Sentinel-2B were successfully launched on June 23, 2015, and March 7, 2017, respectively, their products (Level-2A, which covers the bottom-of-atmosphere reflectance in cartographic geometry; Level-1B, which covers the top-of-atmosphere (TOA) radiance in sensor geometry; and, Level-1C, which covers the TOA reflectance in fixed cartographic geometry) have been widely applied for monitoring land cover changes, agricultural applications, monitoring vegetation and retrieving biophysical parameters, observing coastal zones, monitoring inland water, monitoring glaciers and ice and mapping snow, mapping floods, and management [51–56]. However, these products have not been used for detailed vegetation mapping in arid land environments. Hence, the first objective of this paper is to investigate the performance of the Sentinel-2 MSIL1C product for vegetation mapping in an arid land environment.

Producing any substantial land cover/use maps always requires a specific classification method or ML algorithm. Although many methods and algorithms have been developed for satellite data classification applications, the search for advanced classification methods or algorithms is still a hot filed [57–60]. There are no supper classification methods or algorithms that could universally work at high performance, due to facts that classification performance not only controlled by robustness of adopted methods or algorithms, but also affected by discrimination and identification quality, size and distribution quantity of provided data [61,62]. According to the literatures from RS data classification community, the commonly used ML algorithms are artificial neural networks (ANNs) [63], support vector machine (SVM) [64,65], extreme learning machine (ELM) [66], decision trees (DTs) [67], ensemble methods, such as bagging, adaboost, and RaF [57,68,69], and deep neural networks (DNNs) [70,71]. In most scenarios, these algorithms involve a nominal class variable that has more than two values problem, because the real-world land surface usually recorded by EO sensors with simultaneous discrimination of numerous classes. In general, there are two approaches for addressing this type of problem: 1) adapting the binary algorithm to its multiclass counterpart to deal with multiclass problems directly (e.g., DTs, ANNs, ELMs, RaFs); and, 2) reduce the multiclass problem into several binary subproblems first, then form a multiclass prediction based on the results from several binary classifiers, such as AdaBoost, multiclass SVMs, and ND [72–74]. When compared with the direct approaches, the latter approach is appealing, because it does not involve any changes to the underlying binary algorithm [75]. In particular, a structural risk minimization (SRM) rule-based SVM can successfully work with limited quantities and quality of training samples and it often achieves a higher classification accuracy than linear discriminate analysis (LDA), DTs, ANNs, bagging, AdaBoost, and RaFs [64–66,76–78].

Well-known examples of the second approach are ECOC and pairwise classification, which often result in significant increases in the accuracy [72,75,79]. However, many studies have explicitly proven that ECOC works better than pairwise classification mainly due to its more advanced decoding strategies [80–82]. The ECOC framework consists of two steps: coding and decoding. Popular coding strategies include one-versus-all, one-versus-one, random sparse, binary complete, ternary complete, original and dense random coding, while the most frequently applied decoding designs are Hamming decoding, inverse Hamming decoding, Euclidean decoding, attenuated Euclidean decoding, loss-based decoding, probabilistic decoding, β-density-based distribution decoding, and loss-weighted decoding [72,82]. While the one-versus-one and one-versus-all strategies have been widely adopted in RS data classification, only a few works [83,84] have focused on applications of the ND and its ensemble variants, which have been proven to outperform the direct multiclass, ECOC, and pairwise classification methods while using C4.5 and logistic regression as the base learners [75,85,86]. Additionally, the most recent and more advanced direct multiclass classification methods may also see improved accuracy by interacting with ND and END. Thus, the second objective of this paper is to investigate the performance of the popular ND algorithms, including NDCB, NDDB, NDRPS, NDFC, and their ensemble versions (i.e., ENDCB, ENDDB, ENDRPS, and ENDBC, respectively) by setting C4.5 and bagging [87], AdaBoost [88], an RaF [89], a RoF [90], ExtraTrees [91], and an SVM [92] as the base learners.

The discrimination and identification quality of the provided data is another critical factor that controls the classification performance of the adopted classifier. Over the years, many approaches have been proposed to increase the discrimination and identification ability of the provided data. Among these approaches, structural filtering, MPs, random fields, object-based image analysis (OBIA) and geographic OBIA (GEOBIA), sparse representation (SR), and deep learning (DL)-based contextual information extraction are the most undertaken families of methods [77,93–96]. In the last ten years, mathematical morphology (MM)-based operators, such as MPs, EMPs, APs, and MPs with partial reconstruction (MPPR), have been the most widely accepted approaches in the RS image classification community, mainly due to their advanced and proven performances in contextual information extraction from HR/VHR RS imagery [68,77,93,96,97].

However, the SE sequences or attribute filters (AFs) that are necessarily adopted in the above operators always result in computationally inefficient and redundant high-dimensional features, which may become prohibitively large data processing cases. Additionally, the sequences of SE and AFs, with limited sizes and shapes, cannot match all of the sizes and shapes of the objects in an image; specifically, a single SE is not suitable for an entire image in each operation case [97,98]. MSER-MPs, SPMPs, and multi-resolution segmentation (MRS)-OMPs have been proposed for the spectral-spatial classification of VHR multi/hyperspectral images with the ExtraTrees, ForestPA, and ensemble extremely randomized decision trees (EERDTs) ensemble classifiers in our previous works to overcome such challenges [98,99]. MSER_MPs(M), SPMPs(M), and OMPs(M) were also proposed by considering the mean pixel values within regions, such as MSER objects, superpixels, and MRS objects, to foster effective and efficient spatial FE. The improvements from taking the mean values were clear. Specifically, the size of the regions generated was on a reasonable scale, which was mainly controlled by the spatial resolution and a readily available landscape image [99]. However, as hybrid methods of MPs and OBIA, comparison studies between SPMPs and OBIA, and between OMPs and OBIA were not carried out in our previous works. In addition, as generally known from the OBIA and GEOBIA communities, there are plenty of spectral, statistical, spatial, and geometrical measures of regions (i.e., objects) that can be adopted to further improve the classification accuracy [27,100–102]. Thus, extending the OMPs by considering more advanced object measures is interesting, especially when using Sentinel-2A MSIL1C data for vegetation mapping in large coverage areas in arid land environments, which is the last objective of this paper. In Table 1, we provide the acronym with corresponding full names that are used in this paper.

Table 1. Acronyms with corresponding full names used in this paper.

Acronyms	Full Name	Acronyms	Full Name
AA	Average accuracy	MSIL1C	MultiSpectral Instrument Level-1C
AFs	Attribute filters	MSER-MPs	Maximally stable extremal region-guided MPs
ANNs	Artificial neural networks	ND	Nested dichotomies
AVHRR	Advanced VHR Radiometer	NDBC	ND based on clustering
CBR	Closing by reconstruction	NDCB	ND with class balancing
CVRFR	Classification via RaF regression	NDDB	ND with data balancing
DL	Deep learning	NDFC	ND with further centroid
DNNs	Deep neural networks	NDRPS	ND with random-pair selection

Table 1. *Cont.*

Acronyms	Full Name	Acronyms	Full Name
DTs	Decision trees	OA	Overall accuracy
ECOC	Error-correcting output code	OBIA	Object-based image analysis
EERDTs	Ensemble of ERDTs	OBR	Opening by reconstruction
EL	Ensemble learning	OBPR	Opening by partial reconstruction
ELM	Extreme learning machine	OLI	Operational Land Imager
END	Ensembles of ND	OMPs	Object-guided MPs
ENDBC	Ensemble of NDBC	OMPsM	OMPs with mean values
ENDCB	Ensemble of NDCB	OO	Object-oriented
ENDDB	Ensemble of NDDB	OOBR	Object guided OBR
ENDRPS	Ensemble of NDRPS	PCA	Principal component analysis
END-ERDT	END with ERDT	RaF	Random forest
EOMPs	Extended object-guided MPs	RBF	Radial basis function
ERDT	Extremely randomized DT	ROI	Region of interest
ESA	European Space Agency	RoF	Rotation forest
ETM	Enhanced Thematic Mapper	SE	Structural element
ExtraTrees	Extremely randomized trees	SEOM	ESA's Scientific Exploration of Operational Missions
EVI	Enhanced vegetation index	SNAP	Sentinel Application Platform
GEOBIA	Geographic OBIA	SPOT	Satellite for Observation of Earth
GPS	Global positioning system	SR	Sparse representation
HR	High resolution	SRM	Structural risk minimization
LDA	Linear discriminate analysis	SVM	Support vector machine
LR	Logistic regression	SVM-B	SVM with Bayes optimization
ML	Machine learning	SVM-G	SVM with grid-search optimization
MM	Mathematical morphology	SWIR	Short wave infrared
MPs	Morphological profiles	UA	User accuracy
MPPR	MPs with partial reconstruction	UMD	University of Maryland
MRFs	Markov random fields	TOA	Top-of-atmosphere
MRS	Nulti-resolution segmentation	VHR	Very high resolution
MODIS	Moderate Resolution Imaging Spectroradiometer	VI	Vegetation index
MSI	MultiSpectral Instrument	VNIR	Visible and the near-infrared

2. Materials and Methods

2.1. Materials

2.1.1. Study Region

Our study area is located at the Ili River delta, in the central-western part of the Balkhash Lake basin, in the southeastern part of Kazakhstan (Figure 1a). The Balkhash Lake basin is one of the largest internal drainage areas in the arid and semiarid region in Central Asia; it is located between 72.44°–84.99°E and 42.24°–49.14°N, covering an area of approximately 500,000 km^2 and it is shared by the Republic of Kazakhstan (approximately 60%) and the People's Republic of China (approximately 40%) [103]. Balkhash Lake is the world's fifth-largest inland water reservoir (605 km long and 4–74 km wide), with a volume of 87.7 km^3 and a catchment area of 15,730 km^2 [104]. All of the inflow to the Balkhash Lake is received from the western Tien-Shan and the Dzungarsky Alatau and the runoff from their ridges. The two largest rivers flowing into the lake are the Ili River and Karatal River, accounting for approximately 78% and 15% of the total inflow, respectively [105]. Balkhash Lake and several plentiful wetlands in its inflow deltas are considered to be very sensitive ecosystems, whose existence depends on variable climate conditions and extensive human activities, especially in the form of water abstractions from inflows. During the Soviet era, the inflow waters were largely used for irrigation (mainly for rice crops), industry, the water supply to populated areas, and the fishing industry, which resulted in a significant decrease in the water level and the degradation of the surrounding environments [103,105]. After the collapse of the Soviet Union, most of the social and economic activities in the Balkhash Lake basin rapidly diminished, which causes drastic changes in the land cover/use and broad rehabilitation of the ecosystem. Understanding the effects and responses between such drastic changes and the regional environment is crucial for the sustainable development of this basin, which can only be accomplished by the sustainable monitoring of entire environments. Many efforts have been made in recent decades; however, while most studies have focused on water, e.g., water resource management, water level and surface changes, chemical properties, regional-scale land cover/use changes, ecosystem services, and vegetation activity [106–111], only a few studies have focused on basin-level studies while using RS datasets [104]. In almost all of the above studies, datasets from Landsat, the Advanced Very-High-Resolution Radiometer (AVHRR), and MODIS were mainly used. Hence, it is of interest to use more advanced Sentinel-2A MSIL1C products with more advanced spatial FE and ML techniques for vegetation mapping in this area.

2.1.2. Datasets

- Sentinel-2 data collection and preprocessing

In this study, Sentinel-2A geolocated TOA reflectance (L1C) products were acquired from the Copernicus Open Access Hub (https://scihub.copernicus.eu). We selected a total of six images with zero or near-zero (< 10%) cloud coverage, taken between 25 July and 8 August, 2017. Only the visible bands of blue (band 2), green (band 3), red (band 4), and the near-infrared (band 8) region with a 10 m spatial resolution were used. All of the images of the study region were projected to WGS 84/UTM zone 43N and then mosaicked while using the SNAP (v6.0), which is a free, open source software program that is distributed by the ESA under the GNU General Public License and was founded through the ESA's Scientific Exploration of Operational Missions (SEOM) Program. In Figure 1d, Figure 1e, and Figure 1f, true RGB color images were composited by setting band 4 to red, band 3 to green, and band 2 to blue, respectively, for real-world land surface illustration purposes.

- In situ data collection

In total, 120 valid in situ sites (red dots in Figure 1d) were visited on July 27, 28, 29, and 30, 2017. Specifically, 46 field sites were visited on July 27 in the Bakanas irrigation area (Figure 1e), 23 field sites were visited on July 28 in the Ili River delta region, six field sites were visited on July 29 on the way back from Balkhash city to Bakanas District, and 45 field sites were visited on July 30 in the Bakbakty irrigation area (Figure 1f). For all of the field sites, the coordinates were determined while using a differential global positioning system (GPS) and the Chinese BeiDou navigation system, which has a 2 m positioning accuracy. Additionally, the land cover type among 23 possibilities with 19 vegetation types was recorded (Figure 1), between 10 AM and 6 PM local time. Moreover, field sites are determined at locations with only large and uniform spatial coverage of the same land cover type for a more objective and representative in situ site selection. Figure 2 shows the ground photos of representative land cover types in our study area. According to the collected in situ information and further referring to the high-resolution optical images in Google Earth, 582 regions of interest (ROIs) were selected for model training, validation, and data classification for vegetation mapping. Detailed ROI, training sample, and validation sample information are listed in Table 2.

Figure 2. Ground photos of the land cover types.

Table 2. Details of the land cover types in the training and validation samples in the test datasets.

LC Types	1	2	3	4	5	6	7	8	9	10	11	12	13	14	15	16	17	18	19	20	21	22	23	Total
ROIs	253	39	30	17	30	13	6	7	16	1	4	20	9	33	7	6	2	5	19	15	5	8	37	582
Train	147	230	295	144	325	225	128	19	304	12	92	164	42	277	117	45	27	75	976	171	30	238	475	4558
Test	2794	4366	5600	2726	6180	4269	2428	352	5768	218	1753	3113	794	5256	2226	852	516	1429	18542	3246	560	4523	9021	86532
Total	2941	4596	5895	2870	6505	4494	2556	371	6072	230	1845	3277	836	5533	2343	897	543	1504	19518	3417	590	4761	9496	91090

2.2. Methods

2.2.1. Related Methods

- Ensemble of nested dichotomies

In the area of statistics, ND are a standard technique for solving certain multiclass classification problems with logistic regression (LR). Generally, ND can be represented with a binary tree structure, where the set of classes is recursively split into two subsets until there is only one (Figure 3). In other words, the root node of the ND contains all of the classes that correspond to the multiclass classification problem, and each node contains a single class, which means that, for an *n*-class problem, there are *n* leaf nodes and *n*-1 internal nodes. To build an ND approach based on such a tree structure, we perform the following steps: 1) at each internal node, store the instances pertaining to the classes associated with current node but no other instances; 2) group the classes pertaining to each node into two subsets to ensure that each subset holds the classes that are associated with exactly one of the node's two successor nodes; and, 3) train the binary classifier at each node for the resulting binary class problem [75,85]. If the adopted binary classifier at each node can compute the class probability, the ND can compute class probability in a natural way, which is a convenient feature in real-world applications [112].

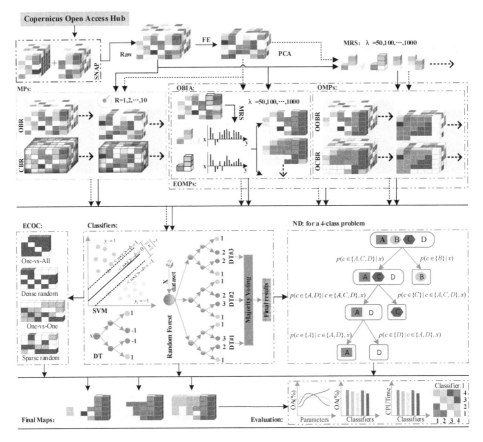

Figure 3. Overall technical flowchart for the methodology.

After an ND approach is built, one critical question is how to combine the probability estimates from individual binary problems to obtain class probability estimates for the original multiclass problem. Multiclass probability estimates can be obtained by simply multiplying the probability estimates returned from the individual binary learners because the individual dichotomies are statistically independent as they are nested. More specifically, let C_{i1} and C_{i2} be the two subsets of classes generated by a split of the set of classes C_i at internal node i of the tree structure, and let $p(c \in C_{i1}|x, c \in C_i)$ and $p(c \in C_{i2}|x, c \in C_i)$ be the conditional probability distributions that are estimated by the binary learner at node i for a given instance x. Subsequently, we can have the estimated class probably distribution for the original multiclass problem by [85]:

$$p(c = C|x) = \prod_{i=1}^{n-1} (I(c \in C_{i1})p(c \in C_{i1}|x, c \in C_i) + I(c \in C_{i2})p(c \in C_{i2}|x, c \in C_i)) \tag{1}$$

where $I(\cdot)$ is the indicator function and the product is over all the internal nodes. Notably, not all of the nodes must be examined to compute the probability for a particular class value, which makes the evaluation of the path to the leaf associated with that class sufficient.

Ever since the basic form of the class subset split criterion was originally proposed by Frank and Kramer [85], many other sophisticated criteria, such as random selection, random-pair selection, clustering, multisubset evaluation, class-balanced-based optimization, data-balanced-based optimization, and genetic algorithm-based optimization, have been proposed and proven to have superior performance on the classification accuracy and model training efficiency, especially with END, which use common EL algorithms such as bagging, boosting, and RaFs [74,75,85,86,112–114].

According to the formation by Frank and Kramer [85], there are $T(c) = (3^c - (2^{c+1} - 1))/2$ possible dichotomies for a c-class problem, which is very large and not ideal for efficient model training. Especially when large amounts of data are readily available, advanced, but computationally inefficient learners (e.g., ANNs and SVMs) are adopted in an ensemble scenario. One simple solution is using random selection dichotomies instead of complete selection dichotomies, which reduces the number of possible dichotomies to $T(c) = (2c - 3) \times T(c - 1)$, where $T(1) = 1$. Briefly, all the distinct dichotomies for a given n-class problem were uniformly sampled with replacement, and the class probability estimates for a given instance x were obtained by averaging the estimates from the individual END members. According to statistical theory regarding EL, reduced numbers of dichotomies are still large enough to ensure that there is a high level of diversity among END members to facilitate the improvement by the ensemble. One drawback of random selection is that it can produce very imbalanced tree structures, which results in a negative effect on the training time of the full model while the number of internal nodes remains the same in any ND for the same number of classes because an unbalanced tree often implies that the internal binary learners are trained on large datasets near the leaves. Dong et al. [75] proposed class-balanced and data-balanced versions of ND, namely, the NDCB and NDDB, respectively, to mitigate the effect of this issue. When compared with NDCB, NDDB can avoid the potential problem from multiclass problems with imbalanced samples. Empirical experiments have shown that NDCB and NDDB have little effect on the accuracy in most cases, but they have great benefits in reducing the time needed for model training, particularly for problems with many classes in ensemble NDCB and NDDB (ENDCB and ENDDB, respectively). The growth function for NDCB is [75]:

$$T(c) = \begin{cases} \frac{1}{2}\binom{c}{c/2}T\left(\frac{c}{2}\right)^2, & \text{if } c \text{ is even} \\ \binom{c}{(c+1)/2}T\left(\frac{c+1}{2}\right)T\left(\frac{c-1}{c}\right), & \text{if } c \text{ is odd} \end{cases} \tag{2}$$

where $T(2) = T(1) = 1$.

When constructing an effective EL system, one is always faced with a dilemma between high classifier diversity and excellent performance, which is hard to satisfy in practice. Unfortunately, the above END methods are deterministic when generating subclass groups that cannot maintain the benefits of high diversity. Additionally, errors that are made by binary classifiers at earlier nodes can be inherently spread to lower nodes of the ND tree and they cannot be easily corrected. For these reasons, it is important to generate the dichotomies in a nondeterministic way to reach the high diversity requirement on the one hand and reduce the number of errors in and the size of the upper nodes of the ND tree on the other hand. NDBC, NDRPS, and their ensemble versions are good examples for this intension [74,112]. However, as compared with NDCB, NDRPS is more direct, easily discovers similar classes, and exhibits a degree of randomness, which leads to more diversity and a higher-performing ensemble. The growth function of the NDRPS was empirically estimated by [112]:

$$T(c) = p(c)T\left(\frac{c}{3}\right)T\left(\frac{2c}{3}\right) \tag{3}$$

where $p(c)$ represents the size of the dichotomies from the base learner and $T(2) = T(1) = 1$.

- Multiresolution segmentation

OBIA is a classic technique in RS image interpretation that integrates the spatial and spectral features and it splits RS images into a set of nonoverlapping homogeneous regions or objects, depending on the segmentation method that was specified. During recent decades, OBIA has gained widespread attention in the RS community, mainly because it can overcome the limitations of pixelwise analysis, such as the neglect of geometric, contextual, and semantic information, particularly in the processing of HR/VHR RS imagery [100–102,115]. Over the years, many image segmentation methods have been proposed and extensively examined while using various RS imagery. Among these methods, MRS is one of the most frequently used methods, which is mainly due to its capability to produce high-quality segments at different scales [69,102,116,117].

MRS is a bottom-up region-merging-based segmentation technique that starts with one-pixel objects and it merges the most similar adjacent pixels or objects provided that the internal heterogeneity of the resulting object does not exceed a user-defined threshold [118]. The heterogeneity measure in eCognition considers the spectral heterogeneity, which allows for multivariant segmentation by adding weights to the image channels, and the shape heterogeneity, which describes the improvement in the shape with regard to the smoothness and compactness. In any OBIA, the segmentation scale determines the average size and number of segments that were produced. Defining an optimal scale segmentation to avoid oversegmentation and undersegmentation issues is always challenging because of the spectral similarity between different objects and landscape complexity in the real world [101,102,119]. For MRS, numerous studies demonstrate the importance of the scale parameter, because it controls the dimensions and the size of the segments, which may directly affect subsequent results [101,117,119]. A successful research result on scale optimization is to combine the local variance (LV) and the rates of change of the LV (ROC-LV) to determine the appropriate segmentation scales for a given resolution [120]. The automated selection of scale parameters is basically an automation of the ESP tool, where the production of a graph is replaced by an iterative procedure that segments an image at the first threshold that occurs in the LV graph. The readers are referred to the original works by [118] and [120] for more detailed information.

2.2.2. Proposed Method

MPs are composed of morphological opening and closing profiles, which consist of an ensemble of OBR and CBR operators. According to the definition of MPs, OBR and CBR operators are connected operators that satisfy the assertion of removing the structures that cannot contain the SE and preserving those structures that can contain the SE [121–123]. While applying such operators with a sequence of SEs of increasing size, one can extract information regarding the contrast and the size of the geometrical structures that are present in the image. Originally, the formulation of the spatial information that was included in the MPs refers to a single-band image; therefore, the direct construction of MPs is not straightforward for multi/hyperspectral images. Several approaches have been considered to overcome this shortcoming [77,93,94,124,125]. Among these approaches, one simple, and yet efficient, approach is to use a few images that contain most of the spectral information that was obtained by some FE method, namely, the EMPs [121]. If we consider the first m principal components that were extracted from the multi/hyperspectral images with principal component analysis (PCA), the EMPs are obtained by stacking all of the MPs that are built on all m components.

According to the definition from MM and our previous works [98,99], the MRS object-guided morphological OBR operators can be obtained by first eroding the input image while using segmented objects (where Θ_S^λ represents the numbers (S) of objects from MRS with scale λ) in the SE approach and by using the result as a marker in geodesic reconstruction by a dilation phase:

$$OOBR(f) = R_f^D\left[f \odot (\exists\Theta_{j,j\in S}^\lambda \in \Theta_S^\lambda)\right] \tag{4}$$

Similarly, we have

$$OCBR(f) = R_f^E\left[f \oplus (\exists\Theta_{j,j\in S}^\lambda \in \Theta_S^\lambda)\right] \tag{5}$$

where the object-guided CBR (OCBR), which was obtained by complementing the image f^C, contains the object-guided OBR (OOBR) with SEs $\exists\Theta_{j,j\in S}^\lambda$ and it complements the resulting procedure:

$$OCBR(f) = R_f^{DC}\left[f^C \odot (\exists\Theta_{j,j\in S}^\lambda \in \Theta_S^\lambda)\right] \tag{6}$$

In MM, the erosion of f by b at any location (x, y) is defined as the minimum value of all the pixels in its neighborhood, denoted by b. In contrast, dilation returns the maximum value of the image in the window that was outlined by b. Subsequently, we can have the following new formations for the erosion and dilation operators:

$$\left[f \odot (\exists\Theta_{j,j\in S}^\lambda \in \Theta_S^\lambda)\right](x, y) = \min_{(s,t)\in\Theta_{j,j\in S}^\lambda} \{f(x+s, y+t)\}$$
$$\left[f \oplus (\exists\Theta_{j,j\in S}^\lambda \in \Theta_S^\lambda)\right](x, y) = \max_{(s,t)\in\Theta_{j,j\in S}^\lambda} \{f(x+s, y+t)\} \tag{7}$$

By substituting Equation (13) into Equations (10) and (12), we have the formations of the OOBR and the OCBR as:

$$OOBR(f) = R_f^D\left[\min_{(s,t)\in\Theta_{j,j\in S}^\lambda} \{f(x+s, y+t)\}\right]$$
$$OCBR(f) = R_f^E\left[\max_{(s,t)\in\Theta_{j,j\in S}^\lambda} \{f(x+s, y+t)\}\right] = R_f^{DC}\left[\min_{(s,t)\in\Theta_{j,j\in S}^\lambda} \{f^C(x+s, y+t)\}\right] \tag{8}$$

If the SEs $\exists\Theta_{j,j\in S}^\lambda$ are specified by MRS objects with a sequence of scale parameter λ, then the MRS object guided morphological profiles (OMPs) of an image f can be defined as:

$$OMPs(f) = \left[OOBR(f)^{(\exists\lambda\in\{\lambda_1^*,\lambda_2^*,...,\lambda_Q^*\})}, OCBR(f)^{(\exists\lambda\in\{\lambda_1^*,\lambda_2^*,...,\lambda_Q^*\})}\right] \tag{9}$$

where $\{\lambda_1^*, \lambda_2^*, \ldots, \lambda_Q^*\}$ represents the sets of Q numbers of the user-specified scale parameter λ.

By further considering the extensively proven performance from object profiles in OO-based image classification, the extended OMPs (EOMPs) can be calculated, as follows:

$$EOMPs(f) = \left[OOBR(f)^{(\exists \lambda \in \{\lambda_1^*, \lambda_2^*, \ldots, \lambda_Q^*\})}, OCBR(f)^{(\exists \lambda \in \{\lambda_1^*, \lambda_2^*, \ldots, \lambda_Q^*\})}, (f)_{OO}^{(\exists \lambda \in \{\lambda_1^*, \lambda_2^*, \ldots, \lambda_Q^*\})} \right] \quad (10)$$

where

$$(f)_{OO}^{(\exists \lambda \in \{\lambda_1^*, \lambda_2^*, \ldots, \lambda_Q^*\})} = \left[\left(O_{Min'}^{\lambda_k^*}, O_{Mean'}^{\lambda_k^*}, O_{Max'}^{\lambda_k^*}, O_{Std'}^{\lambda_k^*}, O_{Roun.'}^{\lambda_k^*}, O_{Comp.'}^{\lambda_k^*}, O_{Asym.'}^{\lambda_k^*}, O_{Rect.'}^{\lambda_k^*}, O_{MeanIn.'}^{\lambda_k^*}, O_{Density'}^{\lambda_k^*}, O_{BorderI.'}^{\lambda_k^*}, O_{ShapeI.'}^{\lambda_k^*}, O_{Elliptic}^{\lambda_k^*} \right) \right]_{k=1,\ldots,Q} \quad (11)$$

represents the collections of 13 object features, including pixel value-based measures, such as the minimum, the maximum, the mean, the standard deviation, and the mean of the inner border, and geometrical measures, such as the roundness ($O_{Roun.}^{\lambda_k^*}$), the compactness ($O_{Comp.}^{\lambda_k^*}$), the asymmetry ($O_{Asym.}^{\lambda_k^*}$), the rectangular fit ($O_{Rect.}^{\lambda_k^*}$), the border index ($O_{BorderI.}^{\lambda_k^*}$), the shape index ($O_{ShapeI.}^{\lambda_k^*}$), and the elliptic fit ($O_{Elliptic}^{\lambda_k^*}$).

Finally, Figure 3 shows the overall technical flowchart for the proposed method.

2.2.3. Experimental Setup

To analyze the performance of the introduced the multiclass classification methods ND and END, state-of-the-art and classic ML algorithms, including C4.5 [87], END with ERDT (END-ERDT) [91], RaFs [89], ExtraTrees [91,98], classification via random forest (CVRaFs) [126], RoFs [90], and an SVM [64], were also applied in direct- or ECOC-based multiclass classification. The considered ECOC methods include one-versus-one (ECOC:1vs1), one-versus-all (ECOC:1vsAll), random correlation (ECOC:RC), dense random (ECOC:DR), sparse random (ECOC:SR), and ordinal (ECOC:Ordinal) methods. Critical tree parameters of C4.5, END-ERDT, RaF, RoF, CVRaF, and ExtraTrees classifiers are set by default, while the ensemble size is set to 100 by default for RaF, RoF, CVRaF, and ExtraTrees. The involved parameters of the radial basis function (RBF) kernel-based SVM were tuned by using Bayes optimization (SVM-B) and 10 by 10 grid-search optimization (SVM-G) [127].

We applied a disk-shaped SE with n = 10 openings and closings by conventional and partial reconstructions to obtain the MPs and MPPR from the four raw bands of MSIL1C and the first three PCA-transformed components, ranging from one to ten with a step-size increment of one. These parameters mean that we obtain $84 = 4 + 4 \times 10 \times 2$ dimensional datasets using four raw bands and $63 = 3 + 3 \times 10 \times 2$ dimensional datasets using the first three PCA-transformed components, which are represented by Raw_MPs, Raw_MPPR, PCA_MPs, and PCA_MPPR in the graphs in the experimental parts. For fair evaluations from dimensionality, we set the MRS segregation scale parameter λ with 10 different values in the FE phase for OMPs and EOMPs. In other words, we obtained $84 = 4 + 4 \times 10 \times 2$ and $63 = 3 + 3 \times 10 \times 2$-dimensional datasets for the raw and PCA-transferred data, respectively, while using OMPs and $524 = 4 + 4 \times 13 \times 10$ and $393 = 3 + 3 \times 13 \times 10$ dimensional OO feature datasets from the raw and PCA-transformed data, respectively. Naturally, there are $604 = (524 - 4) + 84$ and $453 = (393 - 3) + 63$ dimensional datasets for raw and PCA-transferred data, respectively, while using EOMPs.

In the experiment, the average accuracy (AA), the overall accuracy (OA), the CPU running time (CPUTime), and the kappa statistic were used to evaluate the classification performance of all the considered methods. All of the experiments were conducted while using Oatave 5.1.0 on a Windows 10 64-bit system with an Intel Core i7-4790 3.60 GHz CPU and 64 GB of RAM.

3. Results

3.1. Subsection Assessment of the Feature Extractors

3.1.1. Accuracy Evaluation

Figure 4 illustrates the OA values from the ensemble methods, including RaF, ExtraTrees, and END-ERDT while using MPs, MPPR, and EOMP features that were extracted from the raw and PCA-transformed datasets. Each point on the *x*-axis represents the MRS scale sets for OO, OMPs, OMPsM, and EOMPs feature extractors (e.g., 50-500-50 means the scale parameter λ of MRS that starts with 50 and stops at 500 with total 10 steps by step 50), while the *y*-axis representation the OA values. First, the superiority of the proposed FE method EOMPs is obvious when compared with that of the MPs, MPPR, OMPs, and OMPsM, and the superiority of OO as compared with that of the MPs, MPPR, and OMPs. Specifically, the best improvements were achieved by EOMPs across all three classifiers with two datasets (see the dark green lines). Moreover, the superiority of MPPR compared to MPs and OMPs and the superiority of OMPsM when compared to MPPR is clear, which again supports the findings by Liao et al. [97] and Samat et al. [98]. Additionally, the performance of OO and OMPs could actually be limited by setting the segmentation scale parameter λ to very large values. For example, a decreasing trend in the OA values from OMPs can be observed after the starting scale is larger than 100 with 100 or 50 scale steps (see the brown lines).

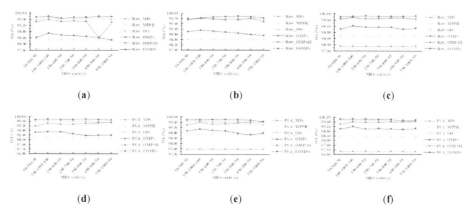

Figure 4. Overall accuracy (OA) values from RaF (**a** and **d**), ExtraTrees (**b** and **e**), and END-ERDT (**c** and **f**) using morphological profiles (MPs) (disk, 1-10-1), MPs with partial reconstruction (MPPR) (disk, 1-10-1), and extended object-guided MPs (EOMP) features extracted from the raw bands.

3.1.2. Visual Evaluation

In Figure 5, a 600 × 800 image patch was selected from the south-central area of the Bakbakty irrigation area (Figure 1f) to show the differences between the OBR, opening by partial reconstruction (OBPR) and object guided OBR (OOBR) operators with different scale parameter settings while using the first raw band. According to the graphs in the first row of Figure 4, the image becomes slightly grayer as the size of the SEs increases in OBR, with most of the small details, such as boundaries between objects, still remain. In contrast, boundaries between different objects become too blurred and indistinguishable as the size of the SEs increases in OPPR; many large objects, such as the urban area in the central-western area that should appear at a certain scale of the area attribute, remain at a low scale, and disk shapes, such as new objects, are created with large SE size after OPPR (see the last image in row 2 of Figure 4). For the proposed EOMPs, the target image becomes slightly grayer as the scale parameter λ, which controls the total number and individual scales of segments, increases. Furthermore, most of the boundaries between the different land cover types remain exactly as in the original, which is mainly due to the EOMPs only filtering the areas within the corresponding boundaries. However, when the segments are too large and are composed of many different objects, the performance of OOBR could be limited by returning profiles of only one object. In addition, different segments could have the same minimum and/or maximum pixel values; as a result, OOBR could return similar, or even the same, profiles for different objects. For example, a very bright and rectangular building in the center of the target image cannot be distinguished from its surroundings when the value of λ is greater than 600 (see the last row of Figure 4).

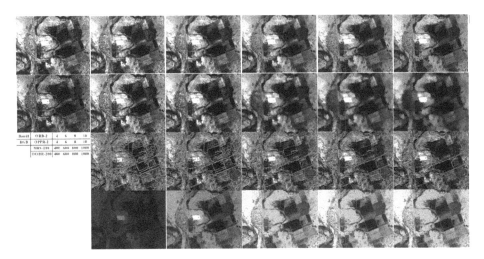

Figure 5. Examples of OBR (row 1), opening by partial reconstruction (OBPR) (row 2), multi-resolution segmentation (MRS) segments (row 3), object-guided OBR (OOBR) (row 4) computed from band 1 (first image at row 1) at the center-bottom of Figure 1f (the numbers in the table in row 3 show the disk sizes in OBR and OBPR and the segmentation scale λ in MRS).

3.2. Evaluation of ND and END

3.2.1. Classification Accuracy

As mentioned in Part 1, the second objective of this paper is to investigate the performance of popular ND algorithms and their ensemble versions. Hence, Figure 6 presents the OA values from various classifiers that were adopted in direct, ND, ECOC, and END multiclass classification frameworks by using all of the considered features.

If we simply compare the OA bars from all of the adopted classification algorithms while using various features in all three multiclass classification framework scenarios, the results of MPPR are superior to those of MPs and OMPs, and the results of OMPs are superior to those of MPPR, OO is superior to MPs and MPPR, EOMPs is superior to all others, and are uniformly shown in almost all of the classification scenarios, which confirms the superiority of our proposed method.

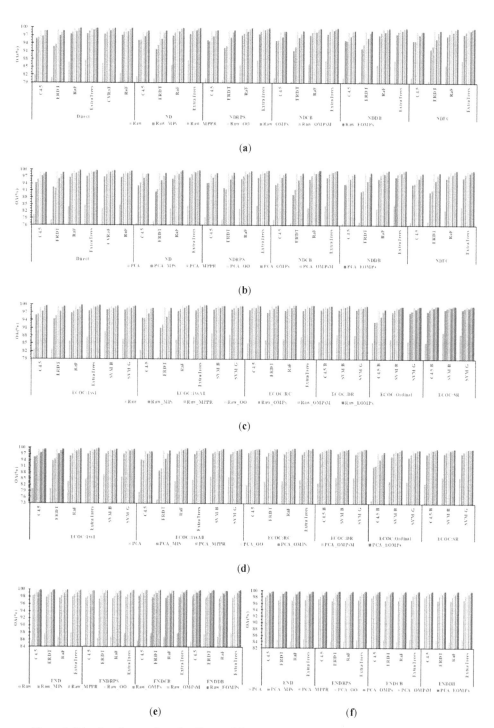

Figure 6. OA values from various classifiers in different multiclass classification frameworks (Direct and ND: a, b; ECOC: c, d; END: e, f) using all the considered features.

When comparing the OA bars in Figures 6a and 6b from C4.5, ERDT, RaF, and ExtraTrees in direct and ND multiclass classification, improvements from ND, NDCB, NDDB, NDRPS, and NDFC over the direct framework is not clear. Interestingly, the performance of the weak, direct multiclass classification algorithm can be reduced in the case of readily available low-dimensional data with low discrimination capability. For instance, the C4.5 classifier reached OA values that were greater than 82% and 80% individually while using the original raw bands and the PCA-transformed datasets, respectively, in the direct multiclass classification framework; moreover, C4.5 in ND, NDCB, NDDB, NDRPS, and NDFC multiclass classification frameworks uniformly reached OA values that were less than 82% and 79% while using the original raw bands and the PCA-transformed datasets, respectively. Similar results can also be found for ERDT not only using the original raw and PCA-transformed datasets, but also using MPs and MPPR features from the original raw and PCA-transformed datasets, whereas the ERDT has proven much weaker than C4.5 [98]. When comparing the OA bars of ensemble classifiers, such as RaF and ExtraTrees, there are no obviously increased or decreased OA values observed for ExtraTrees in the direct and ND, NDCB, NDDB, NDRPS, and NDFC multiclass classification frameworks, but a slightly decreasing trend is shown by RaF in the ND, NDCB, NDDB, NDRPS, and NDFC frameworks while using the original raw and PCA-transformed datasets.

According to the results that are shown in Figures 6c and 6d, there are no obvious increases or decreases in the OA for the same classifiers with different ECOC techniques, except for ERDT and C4.5 in the one vs. all (1 vs. all) and C4.5 in ordinal multiclass classification cases while using original raw and PCA transformed datasets. Additionally, differences in OA values from ECOC techniques using OO and spatial features are smaller and more stable than those from the ND frameworks. Take the C4.5 classifier as an example, 95%–99% and 93%–98% OA value ranges for ND multiclass classification framework becomes into 98%–99.80% and 97.5%–99.80% OA values for ECOC RC. Additionally, more interestingly in comparing with direct and ND frameworks, better OA results can always be reached for weak classifiers (e.g., C4.5, ERDT) in ECOC one vs. one, random correlation, dense random, and sparse random multiclass classification techniques. For instance, a minimum larger than 82% (ECOC 1vs all) and maximum around 86% (ECOC RC, DR and SR) OA values are shown by C4.5 in ECOC frameworks while using original raw bands, while minimum larger than 81% (NDRP) and maximumly larger than 82% (ND) OA values are shown in ND, NDCB, NDDB, NDRP, and NDFC frameworks. On the contrary, when the stronger classifiers, such as RaF, ExtraTree, and SVM are adopted, differences between them in direct, ND, and ECOC frameworks are much smaller, especially from those using high dimensional datasets with high discrimination capabilities. In contrast with RaF and ExtraTrees, better OA values could be reached by SVM in ECOC frameworks while using low dimensional datasets with low discrimination capabilities in the original raw bands and PCA-transformed datasets.

By comparing the results in Figures 6e and 6f with the results in Figures 6a and 6b, we can clearly observe the superiority in the OA values of END, ENDRPS, ENDCB, and ENDDB over ND, NDRPS, NDCB, and NDDB, respectively, which is in accordance with the findings from Frank and Kramer [93], Dong et al. [83], and Rodríguez et al. [94]. Interestingly, the OA values of ERDT in the END, ENDRPS, ENDCB and ENDDB frameworks always reached better OA values than C4.5 and RaF (except for ENDRPS) with the same multiclass classification sets, even when using the original raw bands and PCA-transformed datasets with low discrimination capabilities (see the bars in light blue in Figures 6e and 6f). When better data with high discrimination capabilities are available, the END, ENDRPS, ENDCB, and ENDDB multiclass classification frameworks are capable of reaching better OA values while using weak but simple classifiers (e.g., C4.5 and ERDT) than direct and ECOC when using stronger but more complex classifiers (e.g., RaF, ExtraTrees, and SVM). For example, the OA values for C4.5 and ERDT are approximately 98% larger in the END framework, while the OA values for SVM-B and SVM-G are approximately 97% larger in the ECOC:Ordinal framework while using various spatial features that were extracted from the original raw bands.

In Figure 7, we present the OA curves from the direct and END-based multiclass classification frameworks with incrementally increased ensemble size. The conventional C4.5 and ERDT classifiers are adopted in direct multiclass classification approaches RaF and ExtraTrees, respectively. C4.5, ERDT, RaF, and ExtraTrees are adopted as the base learners in the END, ENDCB, ENDDB, and ENDRPS frameworks. Note that the size of RaF and ExtraTrees are set to 100 in the END, ENDCB, ENDDB, and ENDRPS frameworks. Based on the results, the superiority of ENDCB and ENDDB over END is not obvious in the context of the OA values, as shown in a study by Dong et al. [75]. In contrast, ENDRPS showed the worst results while using the C4.5 and ERDT classifiers. Additionally, the END, ENDCB, and ENDDB frameworks with the ERDT classifier can achieve classification accuracy results that are better than those attained by RaF, by using both the original raw bands and the MPs features that were extracted from raw bands (see the results in Figures 7a and 7e). However, optimum results can be reached by feeding the ExtraTrees to the END, ENDCB, ENDDB, and ENDRPS frameworks. For effects from the ensemble size, increasing the ensemble size beyond 80 does not yield obvious improvements in the OA values for the END, ENDCB, ENDDB, and ENDRPS frameworks with C4.5 and ERDT while using the considered features, while increasing the ensemble size beyond 30 does not yield obvious improvements in the OA values for the END, ENDCB, ENDDB, and ENDRPS frameworks with RaF and ExtraTrees.

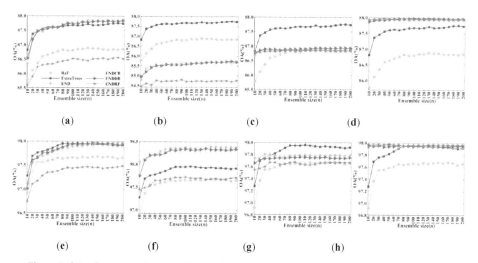

Figure 7. OA values versus the ensemble size of the END, ENDCB, ENDDB, and ENDRPS frameworks with ERDT (**a, e**), C4.5 (**b, f**), RaF (**c, g**), and ExtraTrees (**d, h**) classifiers while using the original raw bands (**a, b, c, d**) and the MPs (**e, f, g, h**).

3.2.2. Computational Efficiency

Computational efficiency is always considered to be another key factor after the classification accuracy when evaluating a classifier's performance. In accordance with Figures 6 and 7, Figure 8 shows the CPUTime (in seconds) in the training phase for various classifiers in different multiclass classification frameworks and using all of the considered features, while Figure 9 shows the results for the END, ENDCB, ENDDB, and ENDRPS frameworks with different ensemble sizes.

When comparing the charts in Figure 8, direct ERDT is at least 10 to 1000 times faster than the C4.5, RaF, ExtraTrees, CVRaF, and RoF classifiers, ExtraTrees is faster than RaF, CVRaF, and RoF, which is in accordance with our previous findings [98]. The extremely fast operability of ERDT is inherently available in the ND, NDCB, NDDB, NDFC multiclass classification frameworks, and in their ensemble versions, as shown in Figures 8e and 8f. Specifically, using ERDT in the ND, NDCB, NDDB, NDFC multiclass classification frameworks is at least 10 times faster than using C4.5 and at least 100 times faster than using RaF and ExtraTrees. It is reasonable that the ensemble size of RaF and ExtraTrees is set to 100 as the default in those frameworks.

In contrast with results from the ECOC frameworks, as shown in Figures 8c and 8d, C4.5, ERDT, RaF, and ExtraTrees in the ND, NDCB, NDDB, NDRPS, and NDFC frameworks are slightly faster than their corresponding frameworks in the ECOC:1vs1, ECOC:1vsAll, and ECOC:RC frameworks. As expected, the worst computational efficiency is shown by the ECOC frameworks with SVM-B and SVM-G parameter optimization techniques. Specifically, SVM-B is 10 times faster than SVM-G, whereas the former is at least 1000 times slower than ERDT in the ND frameworks and at least 100 times slower than ERDT in the END frameworks.

Critical tree parameters, including the minimum leaf size and the maximum depth, are also tuned using Bayes optimization in the ECOC:DR, ECOC:Ordinal, and ECOC:SR frameworks to identify the computational effects from parameter optimization. As shown in Figures 8c and 8d, the computational burden from the parameter tuning process is also severe for C4.5. For instance, the ECOC:SR framework with C4.5 took approximately 1000 seconds of CPUTime on the four original raw bands, while less than 5, 10, and 100 seconds are usual in the direct, ND, NDCB, NDRPS, NDDB frameworks, and their ensemble version frameworks. If we correspondingly look back at the OA results that are shown in Figure 6, obvious improvements in the OA values are not indicated. In other words, the computational complexity that was brought by parameter optimization could be further eliminated in more sophisticated ECOC multiclass classification frameworks without an obvious reduction in the accuracy.

According to the results that are shown in Figure 9, it is clear that the direct classifier ExtraTrees is faster than RaF, and RaF is faster than the END, ENDCB, ENDRPS, ENDDB multiclass classification frameworks while using C4.5, ERDT, RaF, and ExtraTrees as the base learners. Moreover, the computational efficiency of ENDCB and ENDDB over END is also clear, while all of the computational costs of END, ENDCB, ENDRPS, and ENDDB frameworks linearly increase as the ensemble size increases. Interestingly, both the adopted classifier and the ND frameworks can influence the computational efficiency. For instance, the worst computational efficiency is shown by ENDRPS with ERDT while using both the regional raw and MPs datasets (see Figures 9a and 9e), while END with C4.5, RaF and ExtraTrees showed the worst computational efficiency. According to the results that are shown in Figure 7, ENDRPS with ERDT might not be the optimal choice for both accurate and efficient classification with respect to the performance of the END, ENDCB, and ENDDB frameworks.

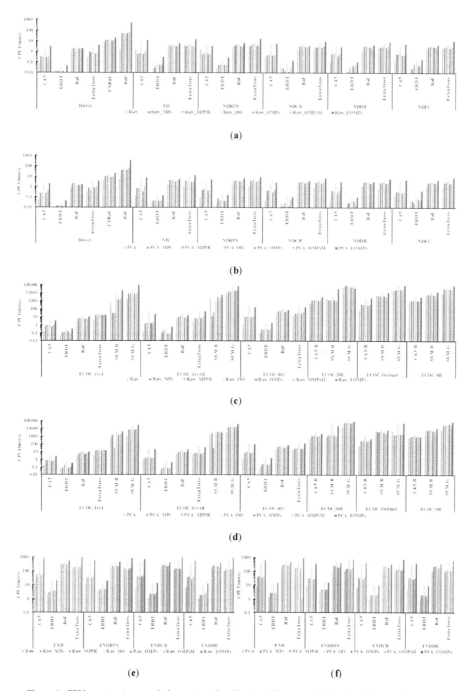

Figure 8. CPU running in seconds for various classifiers in different multiclass classification frameworks (Direct and ND: a, b; ECOC: c, d; END: e, f) using considered features.

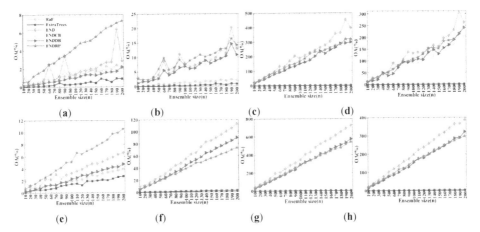

Figure 9. CPU runtime in seconds versus the ensemble sizes of the END, ENDCB, ENDDB, and ENDRPS frameworks with the ERDT (**a**, **e**), C4.5 (**b**, **f**), RaF (**c**, **g**), and ExtraTrees (**d**, **h**) classifiers using the original raw bands (**a**, **b**, **c**, **d**) and features from the MPs (**e**, **f**, **g**, **h**).

3.2.3. Robustness to the Data Dimensionality

Data quality is also a critical factor that controls the classification performance of adopted classifiers, and many approaches can be used to increase the discrimination and identification quality of the provided data by introducing new features. However, increasing the number of data dimensions by introducing new features could limit the training samples large enough to mitigate the Hughes phenomenon on the one hand and increase the computational complexity of feature space splitting-based classifiers (e.g., C4.5, RaF, and RoF) on the other hand. Hence, it is of interest to comparatively investigate the robustness of ND and END to the data dimensionality.

According to the results in Figures 6 and 7, the improved data quality by introducing new features is clear. For various single and ensemble methods, direct and ND-based classifiers, C4.5 is more robust than ERDT to the data dimensionality in the direct, ND, NDCB, NDDB, NDRPS, and NDFC frameworks. For example, ND with ERDT achieves OA values between 92% and 99% after features from MPs, MPPR, OMPs, OMPsM, and EOMPs are introduced, while ND with C4.5 achieves OA values that are between 95% and 99% (see Figure 6a). The ensemble versions of C4.5 and ERDT are less robust than the RaF, RoF, CVRaF, and ExtraTrees, both in direct and various ND. When compared with the results from direct and various ND frameworks, uniformly better robustness to data dimensionality is shown by all of the ECOC frameworks, especially with the RaF, ExtraTrees, and SVM classifiers. Taking the ECOC:RC framework with C4.5 as an example, the OA values range between 95% and 99% for ND with C4.5 and they shrink to a range between 98% and 99% after features from MPs, MPPR, OMPs, OMPsM and EOMPs are introduced.

As expected, the ECOC frameworks with SVM show better robustness to data dimensionality than the ECOC frameworks with C4.5, ERDT, RaF, and ExtraTrees, whereas the SVM is capable of overcoming the Hughes phenomenon that is caused by the data dimensionality with kernel trick [64,65]. When comparing the OA values from various END-based multiclass classification frameworks, it is clear that 1) various END frameworks have better robustness to the data dimensionality than various ND frameworks; 2) differences in the robustness to the data dimensionality between C4.5 and ERDT, C4.5, and RaF, and ERDT and ExtraTrees in various END frameworks are much smaller than those from various ND frameworks; and, 3) similar and even better than ECOC frameworks on the robustness to the data dimensionality can be reached by the END frameworks. For instance, END with C4.5 showed an OA ranging between 98% and approximately 99.8% after various considered features are introduced, while most of the ECOC frameworks with SVM show an OA that ranges between 97% and approximately 99.8%.

As shown in Figure 8, the computational complexity that was brought by the data dimensionality is clear for all classifiers in all of the multiclass classification frameworks. Especially for the C4.5, ERDT, ExtraTrees, RoF, and CVRaF classifiers that adopt feature splits or selection criteria in feature spaces that control the complexity of adopted DTs. For example, a higher computational cost is always shown for ERDT and ExtraTrees in the END frameworks by using Raw_OO (with 524 dimensions) and Raw_EOMPs (with 604 dimensions) features, while a similar and lower computational cost is shown by using Raw_MPs, Raw_MPPR, Raw_OMPs, and Raw_OMPsM features (see Figure 8e). RaF is more robust than the C4.5, ERDT, ExtraTrees, RoF, and CVRaF classifiers to the data dimensionality. From a computational efficiency point of view, the best robustness to the data dimensionality is always shown by ERDT in the direct, ND, ECOC and END frameworks. Additionally, because of the kernel trick, differences in the robustness to the data dimensionality from SVM in ECOC frameworks are smaller than those from the DT-based classifiers that were adopted in the END frameworks.

3.3. Final Vegetation Map

Figure 10 shows the classification map using the proposed method and the considered products to show the superiority of the Sentinel-2 MIL1C products over the MODIS LUCC and GLC30 datasets in arid region. To further compare the findings of END-ERDT capable of reaching the best classification accuracy with a very high computational efficiency, Table 3 reports the classification accuracy values (the user accuracy (UA), AA, OA, and kappa statistics) with CPUTime in seconds for END-ERDT and ECOC:1vsAll with SVM-G optimization.

According to the results in Figure 10, it is apparent that Sentinel-2A MIL1C is better than MODIS LUCC and GLC30 for vegetation diversity mapping in arid regions in Central Asia. Specifically, 19 different vegetation types were recorded by Sentinel-2A MIL1C for our study area, while 15 and eight land cover types were recorded by MODIS LUCC and GLC30 products without specific vegetation taxonomic names. For instance, vegetation species, such as Alhagi sparsifolia, Haloxylon ammodendron, and Artemisia lavandulaefolia are classified as shrubs or herbaceous, while Iris lactea Pall. & Sophora alopecuroides and Sophora alopecuroides are classified into grassland in the MODIS LUCC and GLC30 products. From a vegetation species taxonomy and distribution mapping point of view, the land cover taxonomy classification system might not be appropriate. For example, the vegetation species richness, which is defined as the numbers of different species that are present in a certain study zone, for the Bakanas and Bakbakty irrigation zones that are depicted by blue and green rectangles, respectively, in Figure 10a is four (crops, forest, grass, and shrubs) from the GLC30 product (see Figures 10f and 10g), five (crops, tree, grass, herbaceous, shrubs) from the MODIS LUCC product, and 12 (rice, cloves, wheat, corn, reeds, Alhagi sparsifolia, Carex duriuscula, shrubs, Haloxylon ammodendron, grass, tamarisk, Iris lacteal Pall., and Sophora), and 13 (rice, cloves, wheat, corn, desert steppe, reeds, Alhagi sparsifolia, Carex duriuscula, shrubs, Haloxylon ammodendron, grass, tamarisk, Iris lacteal Pall., and Sophora alopecuroides) from the Sentinel-2 MIL1C classification with END-ERDT while using spectral and spatial features.

Based on the results in Table 3, again, it can be clearly seen that the END-ERDT method is capable of achieving the best results (OA = 99.85%) while using the stacked raw and EOMPs features with the highest model training efficiency (15.20 seconds) with respect to the results from RBF kernel-based SVM-G optimization teaching in the ECOC:1vsAll multiclass classification framework, which confirms the previous findings that END-ERDT could be an alternative to an SVM for generalized classification accuracy, computationally efficient operations, and easy to deploy points of view, especially in the case of sufficient samples with advanced features that are readily available. When the original raw data were adopted, OA values of 87.80% and 88.71% were achieved by the END-ERDT and SVM classifiers, respectively. Furthermore, END-ERDT showed the worst UA of 15.14% for Alhagi sparsifolia, while SVM showed the worst UA values of 2.75% for Alhagi sparsifolia and 1.79% for

Sophora alopecuroides. After the advanced features were included, almost all of the land cover classes were correctly classified with a > 95% UA value by both classifiers, and especially after the OO and EOMPs were included. However, only on the raw data, the END-ERDT model was trained in several to more than ten seconds, the optimum RBF kernel-based SVM model took more than ten thousand seconds.

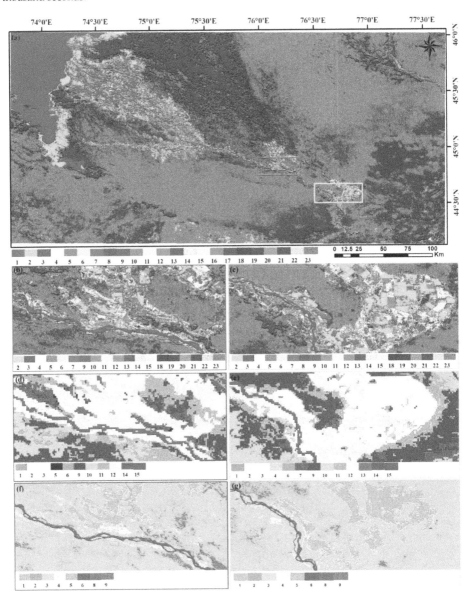

Figure 10. Final vegetation distribution map using Sentinel-2 MSIL1C products with the END-ERDT classifier for our study area (**a**) and subareas (**b**, **c**) and corresponding examples from the 2015 MODIS LUCC products (**d**, **e**) and the 2017 GLC30 (**f**, **g**) products (for the legends, refer to that in Figure 1).

Table 3. Classification accuracy values (user accuracy (UA), average accuracy (AA), OA, and kappa) for the considered methods in the study area.

Class No.	END-ERDT							ECOC:1vsAll (SVM-G)						
	Raw	Raw_MPs	Raw_MPPR	Raw_OMPs	Raw_OMPsM	Raw_OO	Raw_EOMPs	Raw	Raw_MPs	Raw_MPPR	Raw_OMPs	Raw_OMPsM	Raw_OO	Raw_EOMPs
1	54.01	92.27	96.21	96.06	98.50	99.64	99.64	62.53	93.52	93.59	92.95	96.39	97.57	98.85
2	98.12	99.63	99.84	100.00	100.00	100.00	100.00	97.64	99.59	99.86	100.00	99.91	100.00	100.00
3	99.86	99.86	99.88	99.95	99.96	100.00	100.00	99.66	99.52	100.00	99.88	99.86	100.00	100.00
4	63.65	92.99	97.62	98.31	99.85	100.00	100.00	55.58	97.80	97.10	98.86	99.96	100.00	100.00
5	79.79	96.25	98.03	97.59	99.97	100.00	100.00	82.44	96.73	97.98	97.61	99.92	99.92	99.92
6	95.46	99.32	99.88	100.00	100.00	100.00	100.00	97.77	99.11	99.98	100.00	100.00	100.00	100.00
7	89.62	97.65	98.64	99.14	99.14	97.57	98.02	94.28	98.27	99.05	99.63	98.89	97.32	97.32
8	25.28	92.90	97.44	100.00	100.00	100.00	100.00	37.50	99.43	98.86	99.15	99.72	100.00	100.00
9	88.68	98.60	99.03	99.90	99.97	100.00	100.00	88.18	98.75	98.44	99.77	99.69	100.00	100.00
10	15.14	75.23	95.41	94.95	96.79	98.17	100.00	2.75	92.66	97.71	94.95	92.20	100.00	100.00
11	79.52	98.40	99.32	99.54	99.89	100.00	100.00	88.82	98.75	99.83	99.77	99.89	100.00	100.00
12	77.42	94.96	93.61	97.59	98.49	98.33	99.97	79.25	94.47	93.90	94.67	99.16	98.30	98.30
13	58.19	97.23	97.86	99.37	99.24	99.87	99.87	87.41	97.98	96.98	99.12	99.37	99.87	99.87
14	90.56	97.03	98.21	98.99	99.66	99.90	99.96	91.86	98.12	98.82	99.09	99.71	100.00	100.00
15	43.17	91.06	95.46	97.84	99.42	99.28	99.28	37.11	93.89	94.12	98.88	99.46	98.97	99.06
16	53.40	93.54	96.48	97.07	98.71	99.77	99.77	56.57	94.13	95.66	96.71	99.41	99.77	99.77
17	33.33	76.74	95.16	90.50	93.99	95.54	96.12	58.91	91.47	96.32	88.18	92.05	95.16	95.16
18	75.86	97.90	93.98	98.39	99.72	100.00	100.00	73.20	98.25	99.16	98.53	99.51	100.00	100.00
19	99.43	99.94	99.89	99.90	99.98	99.98	100.00	99.36	99.94	99.95	99.92	99.98	100.00	100.00
20	96.12	98.95	99.32	99.63	99.32	99.54	99.51	97.13	98.86	99.17	99.88	99.45	99.54	99.54
21	20.89	76.79	76.79	93.39	97.32	97.50	97.50	1.79	87.50	90.00	90.71	96.61	97.50	97.50
22	89.61	99.31	99.34	99.73	99.96	99.82	99.93	89.94	99.47	99.89	99.78	99.82	99.76	99.89
23	99.92	99.94	99.93	99.93	100.00	100.00	100.00	99.99	100.00	100.00	100.00	100.00	100.00	100.00
AA	70.74	94.20	96.84	98.16	99.13	99.34	99.55	73.03	96.88	97.67	97.74	98.74	99.29	99.36
OA	87.80	97.82	98.62	99.17	99.71	99.75	99.85	88.71	98.42	98.74	99.01	99.60	99.67	99.72
Kappa	0.87	0.98	0.98	0.99	1.00	1.00	1.00	0.88	0.98	0.99	0.99	1.00	1.00	1.00
CPUTime	1.14	2.61	2.16	2.32	3.52	11.22	15.20	11333.50	18625.80	18746.10	20524.90	22582.60	11898.40	62736.90

4. Discussion

For arid land vegetation mapping while using Sentinel-2 MSIL1C image task, the superior performance of the proposed EOMPs over conventional OO, MPs, MPPR, OMPs, and OMPsM is confirmed, both statistically and visually. Additionally, as expected, possible side effects from very large segments could be controlled and even overcome by simply containing the mean pixel values of the objects and the object profiles, such as the compactness, roundness, and shape index. To overcome these potential drawbacks, multiple scale parameter λ should be provided in OOBR and OCBR. On the other hand, those that have been repeatedly proven effective object profiles should also be considered.

With respect to the results from various classifiers in direct, ECOC, ND, and END frameworks, END with ERDT (END-ERDT) always capable of reaching the highest OA values. This finding could be explained by the "diversity" foundation for constructing an effective EL system, which says that "weaker" classifiers (ERDT here) always have a better chance of reaching the trade-off between diversity and accuracy than "stronger" classifies (C4.5 here) [85,128,129]. Additionally, according to statistical theory regarding EL, reduced numbers of dichotomies in ENDCB, ENDDB, and ENDRPS are still large enough to ensure that there is a high level of diversity among END members to facilitate improvement by the ensemble. Hence, investigating the performance of other weak classifiers in END framework will be an interesting topic.

Ensembling randomly generated ND is an effective approach to multiclass classification problems, as proven by the results in Figure 6 and by the works of Frank and Kramer [85]. However, the equal sampling strategy that was adopted in END could limit the classification accuracy by generating a very limited depth of trees that is controlled by the number of classes; moreover, a very unbalanced tree can negatively affect the runtime. To remedy such limitations, NDCB, NDDB, and their ensemble versions (ENDCB and ENDDB, respectively) were proposed by Dong et al. [75]. According to their results, the runtime efficiency of ENDCB and ENDDB were slightly better than that of END in the same cases, and no obvious improvements were observed by setting the ensemble size to a constant value. Hence, it is of interest to comparatively investigate the performances of END, ENDCB, ENDDB, and ENDRPS with various sets of ensemble sizes. Our experiments confirmed that the positive effects from ensemble size are larger for END frameworks with weak classifiers than those with strong classifiers.

In studies that involve RS for biodiversity searches, land cover classification is considered the first-order analysis for species occurrence and mapping [10,23]. In general, coarse-spatial-resolution satellite imagery (e.g., MODIS, TM, and ETM+) and land cover products (e.g., the MODIS land use and cover change (LUCC) and Global Land Cover 30 (GLC30) datasets) are useful in detecting and evaluating ecosystems and habitat structures on a large scale, while HR/VHR satellite imagery products are useful for estimating habitat quality, predicting taxonomic groups, determining species richness, and mapping diversity [130–132]. In arid and semiarid regions, sparsely distributed vegetation species are crucially important in regional ecosystems, but they are easily mixed into dominant land cover types (e.g., bare land) in coarse-resolution satellite imagery. Spectral unmixing and subpixel mapping techniques could eventually solve these problems; however, vegetation species diversity mapping at a fine scale using coarse-resolution satellite images from MODIS, TM, ETM+, and OLI sensors is still quite challenging. For example, spectral unmixing can determine the fractions of classes within mixed pixels, but it fails to predict the spatial location. Our experiment also showed that the Sentinel-2 MIL1C products proved to be a more valuable data source than MODIS LUCC and GLC30 datasets for arid land vegetation mapping. Hence, Sentinel-2 products with 10m, 20m, and 60m spatial resolution, 13 bands spanning from the visible and the near-infrared (VNIR) to the short-wave infrared (SWIR) portion of the spectrum 12 spectral bands, and with a five-day revisit time over land and coastal areas, are better choice than MODIS, TM, ETM+, and OLI for arid land vegetation mapping.

Based on this work, we also envisage future perspectives. Further calculating more advanced vegetation species diversity indices, such as the spectral variation hypothesis (SVH), alpha-diversity, and beta-diversity, to show the superiority of Sentinel-2 MSIL1C images over MODIS and Landsat images should be an interesting future direction, especially at large regional or national scales.

Since END-ERDT showed a state-of-the-art classification performance, statistical and more empirical experiments should both also be conducted. Finally, we will deploy the END-ERDT method on a Spark platform to support big data processing to facilitate its application.

5. Conclusions

Sentinel-2 MSIL1C images of the Ili River delta region of Kazakhstan were classified while using spectral and EOMPs to investigate the performance of the Sentinel-2A MSIL1C products for vegetation mapping in an arid land environment with respect to land cover products from MODIS and Landsat and to answer the question of "is ND and END are superior to state-of-the-art direct and ECOC-based-multiclass classification approaches?" and an accurate classification purposes.

According to the results, several conclusions can be drawn. First and foremost, the proposed EOMP features are better than all of the features, while the OO features are better than the spatial features from the MPs, MPPR, OMPs, and OMPsM for Sentinel-2 MSIL1C image classification. Furthermore, some previous findings of the ND, NDCB, NDDB, NDRPS, and NDFC frameworks showed superiority to direct multiclass classification, and the ECOC approaches are arguably useful in the Sentinel-2 MSIL1C image classification task. This finding can be explained by the fact that the final classification performance is controlled not only by the robustness of the adopted classifier but also by the discrimination capable of providing data. Additionally, the superiority of the END, ENDRPS, ENDCB, and ENDDB frameworks over the ND, NDCB, NDDB, and NDRPS frameworks is confirmed, and one can obtain compatible and even better OA results than the direct and ECOC frameworks by using weak and simple classifiers in the END, ENDRPS, ENDCB, and ENDDB frameworks. For example, END-ERDT can be an alternative to RBF kernel-based SVM in the ECOC framework from the generalized classification accuracy, computationally efficient model training, and easy deployment points of view. Finally, from both greater numbers of species identification and a high classification accuracy point of view, the Sentinel-2A MSIL1C product is more suitable than the global land cover products that are generated from MODIS and Landsat imagery for arid-land vegetation species mapping.

Author Contributions: Conceptualization, A.S. and P.D.; Methodology, A.S.; Situ data collection: A.S., Y.G., L.M. and G.I.; Sentinel 2 data collection and processing: A.S.; Land cover type check using VHR image from Google Earth: C.L.; Original draft preparation: A.S.; Review and editing: A.S., N.Y., P.D., and S.L.; Project admission: A.S. and J.A.; Funding: A.S. and J.A.

Funding: This work was partially supported by the National Natural Science Foundation of China (grant nos. U1603242, 41601440), the Youth Innovation Promotion Association Foundation of the Chinese Academy of Sciences (2018476), and the West Light Foundation of the Chinese Academy of Sciences (2016-QNXZ-B-11).

Acknowledgments: The authors would like to acknowledge the use of free open-access Copernicus Sentinel-2 data (https://scihub.copernicus.eu) and the use of SNAP-ESA Sentinel Application Platform v6.0.0 (http://step.esa.int).

Conflicts of Interest: The authors declare no conflict of interest.

References

1. Chen, Z.; Elvidge, C.D.; Groeneveld, D.P. Monitoring seasonal dynamics of arid land vegetation using AVIRIS data. *Remote Sens. Environ.* **1998**, *65*, 255–266. [CrossRef]
2. Clark, J.S.; Bell, D.M.; Kwit, M.C.; Zhu, K. Competition-interaction landscapes for the joint response of forests to climate change. *Glob. Chang. Biol.* **2014**, *20*, 1979–1991. [CrossRef] [PubMed]
3. Wu, D.; Zhao, X.; Liang, S.; Zhou, T.; Huang, K.; Tang, B.; Zhao, W. Time-lag effects of global vegetation responses to climate change. *Glob. Chang. Biol.* **2015**, *21*, 3520–3531. [CrossRef] [PubMed]
4. Olefeldt, D.; Goswami, S.; Grosse, G.; Hayes, D.; Hugelius, G.; Kuhry, P.; McGuire, A.D.; Romanovsky, V.E.; Sannel, A.B.K.; Schuur, E.A.G.; et al. Circumpolar distribution and carbon storage of thermokarst landscapes. *Nat. Commun.* **2016**, *7*, 13043. [CrossRef]

5. Fleischer, E.; Khashimov, I.; Hölzel, N.; Klemm, O. Carbon exchange fluxes over peatlands in Western Siberia: Possible feedback between land-use change and climate change. *Sci. Total Environ.* **2016**, *545*, 424–433. [CrossRef] [PubMed]

6. Tian, H.; Cao, C.; Chen, W.; Bao, S.; Yang, B.; Myneni, R.B. Response of vegetation activity dynamic to climatic change and ecological restoration programs in Inner Mongolia from 2000 to 2012. *Ecol. Eng.* **2015**, *82*, 276–289. [CrossRef]

7. Jiang, L.; Bao, A.; Guo, H.; Ndayisaba, F. Vegetation dynamics and responses to climate change and human activities in Central Asia. *Sci. Total Environ.* **2017**, *599*, 967–980. [CrossRef] [PubMed]

8. Zhang, Y.; Zhang, C.; Wang, Z.; Chen, Y.; Gang, C.; An, R.; Li, J. Vegetation dynamics and its driving forces from climate change and human activities in the Three-River Source Region, China from 1982 to 2012. *Sci. Total Environ.* **2016**, *563*, 210–220. [CrossRef]

9. Kerr, J.T.; Ostrovsky, M. From space to species: Ecological applications for remote sensing. *Trends Ecol. Evol.* **2003**, *18*, 299–305. [CrossRef]

10. Turner, W.; Spector, S.; Gardiner, N.; Fladeland, M.; Sterling, E.; Steininger, M. Remote sensing for biodiversity science and conservation. *Trends Ecol. Evol.* **2003**, *18*, 306–314. [CrossRef]

11. Madonsela, S.; Cho, M.A.; Ramoelo, A.; Mutanga, O. Remote sensing of species diversity using Landsat 8 spectral variables. *ISPRS J. Photogramm. Remote Sens.* **2017**, *133*, 116–127. [CrossRef]

12. Gould, W. Remote sensing of vegetation, plant species richness, and regional biodiversity hotspots. *Ecol. Appl.* **2000**, *10*, 1861–1870. [CrossRef]

13. Barbier, N.; Couteron, P.; Lejoly, J.; Deblauwe, V.; Lejeune, O. Self-organized vegetation patterning as a fingerprint of climate and human impact on semi-arid ecosystems. *J. Ecol.* **2006**, *94*, 537–547. [CrossRef]

14. Zimmermann, N.E.; Edwards, T.C.; Moisen, G.G.; Frescino, T.S.; Blackard, J.A. Remote sensing-based predictors improve distribution models of rare, early successional and broadleaf tree species in Utah. *J. Appl. Ecol.* **2007**, *44*, 1057–1067. [CrossRef] [PubMed]

15. Xie, Y.; Sha, Z.; Yu, M. Remote sensing imagery in vegetation mapping: A review. *J. Plant Ecol.* **2008**, *1*, 9–23. [CrossRef]

16. Gaitan, J.J.; Oliva, G.E.; Bran, D.E.; Maestre, F.T.; Aguiar, M.R.; Jobbagy, E.G.; Buono, G.G.; Ferrante, D.; Nakamatsu, V.B.; Ciari, G.; et al. Vegetation structure is as important as climate for explaining ecosystem function across Patagonian rangelands. *J. Ecol.* **2014**, *102*, 1419–1428. [CrossRef]

17. Waltari, E.; Schroeder, R.; McDonald, K.; Anderson, R.P.; Carnaval, A. Bioclimatic variables derived from remote sensing: Assessment and application for species distribution modelling. *Methods Ecol. Evol.* **2014**, *5*, 1033–1042. [CrossRef]

18. Tian, F.; Brandt, M.; Liu, Y.Y.; Verger, A.; Tagesson, T.; Diouf, A.A.; Rasmussen, K.; Mbow, C.; Wang, Y.; Fensholt, R. Remote sensing of vegetation dynamics in drylands: Evaluating vegetation optical depth (VOD) using AVHRR NDVI and in situ green biomass data over West African Sahel. *Remote Sens. Environ.* **2016**, *177*, 265–276. [CrossRef]

19. Mildrexler, D.J.; Zhao, M.; Running, S.W. Testing a MODIS global disturbance index across North America. *Remote Sens. Environ.* **2009**, *113*, 2103–2117. [CrossRef]

20. Pettorelli, N.; Laurance, W.F.; O'Brien, T.G.; Wegmann, M.; Nagendra, H.; Turner, W. Satellite remote sensing for applied ecologists: Opportunities and challenges. *J. Appl. Ecol.* **2014**, *51*, 839–848. [CrossRef]

21. Sulla-Menashe, D.; Kennedy, R.E.; Yang, Z.; Braaten, J.; Krankina, O.N.; Friedl, M.A. Detecting forest disturbance in the Pacific Northwest from MODIS time series using temporal segmentation. *Remote Sens. Environ.* **2014**, *151*, 114–123. [CrossRef]

22. McDowell, N.G.; Coops, N.C.; Beck, P.S.; Chambers, J.Q.; Gangodagamage, C.; Hicke, J.A.; Huang, C.Y.; Kennedy, R.; Krofcheck, D.J.; Litvak, M.; et al. Global satellite monitoring of climate-induced vegetation disturbances. *Trends Plant Sci.* **2015**, *20*, 114–123. [CrossRef] [PubMed]

23. Rhodes, C.J.; Henrys, P.; Siriwardena, G.M.; Whittingham, M.J.; Norton, L.R. The relative value of field survey and remote sensing for biodiversity assessment. *Methods Ecol. Evol.* **2015**, *6*, 772–781. [CrossRef]

24. Assal, T.J.; Anderson, P.J.; Sibold, J. Spatial and temporal trends of drought effects in a heterogeneous semi-arid forest ecosystem. *For. Ecol. Manag.* **2016**, *365*, 137–151. [CrossRef]

25. Harvey, K.R.; Hill, G.J.E. Vegetation mapping of a tropical freshwater swamp in the Northern Territory, Australia: A comparison of aerial photography, Landsat TM and SPOT satellite imagery. *Int. J. Remote Sens.* **2001**, *22*, 2911–2925. [CrossRef]

26. Brown, M.E.; Pinzón, J.E.; Didan, K.; Morisette, J.T.; Tucker, C.J. Evaluation of the consistency of long-term NDVI time series derived from AVHRR, SPOT-vegetation, SeaWiFS, MODIS, and Landsat ETM+ sensors. *IEEE Trans. Geosci. Remote Sens.* **2006**, *44*, 1787–1793. [CrossRef]

27. Vieira, M.A.; Formaggio, A.R.; Rennó, C.D.; Atzberger, C.; Aguiar, D.A.; Mello, M.P. Object based image analysis and data mining applied to a remotely sensed Landsat time-series to map sugarcane over large areas. *Remote Sens. Environ.* **2012**, *123*, 553–562. [CrossRef]

28. Dubovyk, O.; Landmann, T.; Erasmus, B.F.; Tewes, A.; Schellberg, J. Monitoring vegetation dynamics with medium resolution MODIS-EVI time series at sub-regional scale in southern Africa. *Int. J. Appl. Earth Obs. Geoinf.* **2015**, *38*, 175–183. [CrossRef]

29. Anchang, J.Y.; Ananga, E.O.; Pu, R. An efficient unsupervised index based approach for mapping urban vegetation from IKONOS imagery. *Int. J. Appl. Earth Obs. Geoinf.* **2016**, *50*, 211–220. [CrossRef]

30. Su, Y.; Guo, Q.; Fry, D.L.; Collins, B.M.; Kelly, M.; Flanagan, J.P.; Battles, J.J. A vegetation mapping strategy for conifer forests by combining airborne LiDAR data and aerial imagery. *Can. J. Remote Sens.* **2016**, *42*, 15. [CrossRef]

31. da Silveira, H.L.F.; Galvão, L.S.; Sanches, I.D.A.; de Sá, I.B.; Taura, T.A. Use of MSI/Sentinel-2 and airborne LiDAR data for mapping vegetation and studying the relationships with soil attributes in the Brazilian semi-arid region. *Int. J. Appl. Earth Obs. Geoinf.* **2018**, *73*, 179–190. [CrossRef]

32. Vrieling, A.; Meroni, M.; Darvishzadeh, R.; Skidmore, A.K.; Wang, T.; Zurita-Milla, R.; Oosterbeek, K.; O'Connor, B.; Paganini, M. Vegetation phenology from Sentinel-2 and field cameras for a Dutch barrier island. *Remote Sens. Environ.* **2018**, *215*, 517–529. [CrossRef]

33. Saveraid, E.H.; Debinski, D.M.; Kindscher, K.; Jakubauskas, M.E. A comparison of satellite data and landscape variables in predicting bird species occurrences in the Greater Yellowstone Ecosystem, USA. *Landsc. Ecol.* **2001**, *16*, 71–83. [CrossRef]

34. Giri, C.; Ochieng, E.; Tieszen, L.L.; Zhu, Z.; Singh, A.; Loveland, T.; Masek, J.; Duke, N. Status and distribution of mangrove forests of the world using earth observation satellite data. *Glob. Ecol. Biogeogr.* **2011**, *20*, 154–159. [CrossRef]

35. Kachelriess, D.; Wegmann, M.; Gollock, M.; Pettorelli, N. The application of remote sensing for marine protected area management. *Ecol. Indic.* **2014**, *36*, 169–177. [CrossRef]

36. Hansen, M.C.; DeFries, R.S.; Townshend, J.R.; Sohlberg, R. Global land cover classification at 1 km spatial resolution using a classification tree approach. *Int. J. Remote Sens.* **2000**, *21*, 1331–1364. [CrossRef]

37. Bartholome, E.; Belward, A.S. GLC2000: A new approach to global land cover mapping from Earth observation data. *Int. J. Remote Sens.* **2005**, *26*, 1959–1977. [CrossRef]

38. Tateishi, R.; Uriyangqai, B.; Al-Bilbisi, H.; Ghar, M.A.; Tsend-Ayush, J.; Kobayashi, T.; Kasimu, A.; Hoan, N.T.; Shalaby, A.; Alsaaideh, B.; et al. Production of global land cover data–GLCNMO. *Int. J. Digit. Earth* **2011**, *4*, 22–49. [CrossRef]

39. Friedl, M.A.; Sulla-Menashe, D.; Tan, B.; Schneider, A.; Ramankutty, N.; Sibley, A.; Huang, X. MODIS Collection 5 global land cover: Algorithm refinements and characterization of new datasets. *Remote Sens. Environ.* **2010**, *114*, 168–182. [CrossRef]

40. Arino, O.; Perez, R.; Ramos Perez, J.J.; Kalogirou, V.; Bontemps, S.; Defourny, P.; Van Bogaert, E. Global land cover map for 2009, European Space Agency (ESA) & Université catholique de Louvain (UCL), PANGAEA. 2012. Available online: https://doi.pangaea.de/10.1594/PANGAEA.787668 (accessed on 30 July 2019).

41. Chen, J.; Liao, A.; Cao, X.; Chen, L.; Chen, X.; He, C.; Han, G.; Peng, S.; Lu, M.; Zhang, W. Global land cover mapping at 30 m resolution: A POK-based operational approach. *ISPRS J. Photogramm. Remote Sens.* **2015**, *103*, 7–27. [CrossRef]

42. Hansen, M.C.; Reed, B. A comparison of the IGBP DISCover and University of Maryland 1 km global land cover products. *Int. J. Remote Sens.* **2000**, *21*, 1365–1373. [CrossRef]

43. Myneni, R.B.; Hoffman, S.; Knyazikhin, Y.; Privette, J.L.; Glassy, J.; Tian, Y.; Wang, Y.; Song, X.; Zhang, Y.; Smith, G.R.; et al. Global products of vegetation leaf area and fraction absorbed PAR from year one of MODIS data. *Remote Sens. Environ.* **2002**, *83*, 214–231. [CrossRef]

44. Hansen, M.C.; DeFries, R.S.; Townshend, J.R.G.; Carroll, M.; Dimiceli, C.; Sohlberg, R.A. Global percent tree cover at a spatial resolution of 500 meters: First results of the MODIS vegetation continuous fields algorithm. *Earth Interact.* **2003**, *7*, 1–15. [CrossRef]

45. Ganguly, S.; Friedl, M.A.; Tan, B.; Zhang, X.; Verma, M. Land surface phenology from MODIS: Characterization of the Collection 5 global land cover dynamics product. *Remote Sens. Environ.* **2010**, *114*, 1805–1816. [CrossRef]
46. Gong, P.; Wang, J.; Yu, L.; Zhao, Y.; Zhao, Y.; Liang, L.; Niu, Z.; Huang, X.; Fu, H.; Liu, S.; et al. Finer resolution observation and monitoring of global land cover: First mapping results with Landsat TM and ETM+ data. *Int. J. Remote Sens.* **2013**, *34*, 2607–2654. [CrossRef]
47. Tuanmu, M.N.; Jetz, W. A global 1-km consensus land-cover product for biodiversity and ecosystem modelling. *Glob. Ecol. Biogeogr.* **2014**, *23*, 1031–1045. [CrossRef]
48. Zhang, H.K.; Roy, D.P. Using the 500 m MODIS land cover product to derive a consistent continental scale 30 m Landsat land cover classification. *Remote Sens. Environ.* **2017**, *197*, 15–34. [CrossRef]
49. Yu, Q.; Hu, Q.; van Vliet, J.; Verburg, P.H.; Wu, W. GlobeLand30 shows little cropland area loss but greater fragmentation in China. *Int. J. Appl. Earth Obs. Geoinf.* **2018**, *66*, 37–45. [CrossRef]
50. Drusch, M.; Del Bello, U.; Carlier, S.; Colin, O.; Fernandez, V.; Gascon, F.; Hoersch, B.; Isola, C.; Laberinti, P.; Meygret, A.; et al. Sentinel-2: ESA's optical high-resolution mission for GMES operational services. *Remote Sens. Environ.* **2012**, *120*, 25–36. [CrossRef]
51. Frampton, W.J.; Dash, J.; Watmough, G.; Milton, E.J. Evaluating the capabilities of Sentinel-2 for quantitative estimation of biophysical variables in vegetation. *ISPRS J. Photogramm. Remote Sens.* **2013**, *82*, 83–92. [CrossRef]
52. Verrelst, J.; Rivera, J.P.; Leonenko, G.; Alonso, L.; Moreno, J. Optimizing LUT-based RTM inversion for semiautomatic mapping of crop biophysical parameters from Sentinel-2 and-3 data: Role of cost functions. *IEEE Trans. Geosci. Remote Sens.* **2014**, *52*, 257–269. [CrossRef]
53. Kääb, A.; Winsvold, S.H.; Altena, B.; Nuth, C.; Nagler, T.; Wuite, J. Glacier remote sensing using Sentinel-2. part I: Radiometric and geometric performance, and application to ice velocity. *Remote Sens.* **2016**, *8*, 598.
54. Novelli, A.; Aguilar, M.A.; Nemmaoui, A.; Aguilar, F.J.; Tarantino, E. Performance evaluation of object based greenhouse detection from Sentinel-2 MSI and Landsat 8 OLI data: A case study from Almería (Spain). *Int. J. Appl. Earth Obs. Geoinf.* **2016**, *52*, 403–411. [CrossRef]
55. Belgiu, M.; Csillik, O. Sentinel-2 cropland mapping using pixel-based and object-based time-weighted dynamic time warping analysis. *Remote Sens. Environ.* **2018**, *204*, 509–523. [CrossRef]
56. Griffiths, P.; Nendel, C.; Hostert, P. Intra-annual reflectance composites from Sentinel-2 and Landsat for national-scale crop and land cover mapping. *Remote Sens. Environ.* **2019**, *220*, 135–151. [CrossRef]
57. Chan, J.C.W.; Paelinckx, D. Evaluation of Random Forest and Adaboost tree-based ensemble classification and spectral band selection for ecotope mapping using airborne hyperspectral imagery. *Remote Sens. Environ.* **2008**, *112*, 2999–3011. [CrossRef]
58. Naidoo, L.; Cho, M.A.; Mathieu, R.; Asner, G. Classification of savanna tree species, in the Greater Kruger National Park region, by integrating hyperspectral and LiDAR data in a Random Forest data mining environment. *ISPRS J. Photogramm. Remote Sens.* **2012**, *69*, 167–179. [CrossRef]
59. Kumar, P.; Gupta, D.K.; Mishra, V.N.; Prasad, R. Comparison of support vector machine, artificial neural network, and spectral angle mapper algorithms for crop classification using LISS IV data. *Int. J. Remote Sens.* **2015**, *36*, 1604–1617. [CrossRef]
60. Omer, G.; Mutanga, O.; Abdel-Rahman, E.M.; Adam, E. Performance of support vector machines and artificial neural network for mapping endangered tree species using WorldView-2 data in Dukuduku forest, South Africa. *IEEE J. Sel. Top. Appl. Earth Obs. Remote Sens.* **2015**, *8*, 4825–4840. [CrossRef]
61. Olofsson, P.; Foody, G.M.; Herold, M.; Stehman, S.V.; Woodcock, C.E.; Wulder, M.A. Good practices for estimating area and assessing accuracy of land change. *Remote Sens. Environ.* **2014**, *148*, 42–57. [CrossRef]
62. Samat, A.; Li, J.; Liu, S.; Du, P.; Miao, Z.; Luo, J. Improved hyperspectral image classification by active learning using pre-designed mixed pixels. *Pattern Recognit.* **2016**, *51*, 43–58. [CrossRef]
63. Mas, J.F.; Flores, J.J. The application of artificial neural networks to the analysis of remotely sensed data. *Int. J. Remote Sens.* **2008**, *29*, 617–663. [CrossRef]
64. Mountrakis, G.; Im, J.; Ogole, C. Support vector machines in remote sensing: A review. *ISPRS J. Photogramm. Remote Sens.* **2011**, *66*, 247–259. [CrossRef]
65. Maulik, U.; Chakraborty, D. Remote sensing image classification: A survey of support-vector-machine-based advanced techniques. *IEEE Geosci. Remote Sens. Mag.* **2017**, *5*, 33–52. [CrossRef]
66. Samat, A.; Du, P.; Liu, S.; Li, J.; Cheng, L. E2LMs: Ensemble Extreme Learning Machines for Hyperspectral Image Classification. *IEEE J. Sel. Top. Appl. Earth Obs. Remote Sens.* **2014**, *7*, 1060–1069. [CrossRef]

67. Xu, M.; Watanachaturaporn, P.; Varshney, P.K.; Arora, M.K. Decision tree regression for soft classification of remote sensing data. *Remote Sens. Environ.* **2005**, *97*, 322–336. [CrossRef]
68. Du, P.; Samat, A.; Waske, B.; Liu, S.; Li, Z. Random forest and rotation forest for fully polarized SAR image classification using polarimetric and spatial features. *ISPRS J. Photogramm. Remote Sens.* **2015**, *105*, 38–53. [CrossRef]
69. Belgiu, M.; Drăguţ, L. Random forest in remote sensing: A review of applications and future directions. *ISPRS J. Photogramm. Remote Sens.* **2016**, *114*, 24–31. [CrossRef]
70. Zhang, L.; Zhang, L.; Du, B. Deep learning for remote sensing data: A technical tutorial on the state of the art. *IEEE Geosci. Remote Sens. Mag.* **2016**, *4*, 22–40. [CrossRef]
71. Zhu, X.X.; Tuia, D.; Mou, L.; Xia, G.S.; Zhang, L.; Xu, F.; Fraundorfer, F. Deep learning in remote sensing: A comprehensive review and list of resources. *IEEE Geosci. Remote Sens. Mag.* **2017**, *5*, 8–36. [CrossRef]
72. Dietterich, T.G.; Bakiri, G. Solving multiclass learning problems via error-correcting output codes. *J. Artif. Intell. Res.* **1994**, *2*, 263–286. [CrossRef]
73. Allwein, E.L.; Schapire, R.E.; Singer, Y. Reducing multiclass to binary: A unifying approach for margin classifiers. *J. Mach. Learn. Res.* **2000**, *1*, 113–141.
74. Duarte-Villaseñor, M.M.; Carrasco-Ochoa, J.A.; Martínez-Trinidad, J.F.; Flores-Garrido, M. Nested dichotomies based on clustering. In *Iberoamerican Congress on Pattern Recognition*; Springer: Berlin/Heidelberg, Germany, September 2012; pp. 162–169.
75. Dong, L.; Frank, E.; Kramer, S. Ensembles of balanced nested dichotomies for multi-class problems. In *European Conference on Principles of Data Mining and Knowledge Discovery*; Springer: Berlin/Heidelberg, Germany, October 2005; pp. 84–95.
76. Foody, G.M.; Mathur, A. A relative evaluation of multiclass image classification by support vector machines. *IEEE Trans. Geosci. Remote Sens.* **2004**, *42*, 1335–1343. [CrossRef]
77. Plaza, A.; Benediktsson, J.A.; Boardman, J.W.; Brazile, J.; Bruzzone, L.; Camps-Valls, G.; Chanussot, J.; Fauvel, M.; Gamba, P.; Gualtieri, A.; et al. Recent advances in techniques for hyperspectral image processing. *Remote Sens. Environ.* **2009**, *113*, S110–S122. [CrossRef]
78. Shao, Y.; Lunetta, R.S. Comparison of support vector machine, neural network, and CART algorithms for the land-cover classification using limited training data points. *ISPRS J. Photogramm. Remote Sens.* **2012**, *70*, 78–87. [CrossRef]
79. Hüllermeier, E.; Vanderlooy, S. Combining predictions in pairwise classification: An optimal adaptive voting strategy and its relation to weighted voting. *Pattern Recognit.* **2010**, *43*, 128–142. [CrossRef]
80. Passerini, A.; Pontil, M.; Frasconi, P. New results on error correcting output codes of kernel machines. *IEEE Trans. Neural Netw.* **2004**, *15*, 45–54. [CrossRef]
81. Pujol, O.; Radeva, P.; Vitria, J. Discriminant ECOC: A heuristic method for application dependent design of error correcting output codes. *IEEE Trans. Pattern Anal. Mach. Intell.* **2006**, *28*, 1007–1012. [CrossRef]
82. Escalera, S.; Pujol, O.; Radeva, P. On the decoding process in ternary error-correcting output codes. *IEEE Trans. Pattern Anal. Mach. Intell.* **2010**, *32*, 120–134. [CrossRef]
83. Pal, M. Class decomposition Approaches for land cover classification: A comparative study. In Proceedings of the IEEE International Geoscience and Remote Sensing Symposium, IGARSS 2006, Denver, CO, USA, 31 July–4 August 2006; pp. 2731–2733.
84. Mera, D.; Fernández-Delgado, M.; Cotos, J.M.; Viqueira, J.R.R.; Barro, S. Comparison of a massive and diverse collection of ensembles and other classifiers for oil spill detection in sar satellite images. *Neural Comput. Appl.* **2017**, *28*, 1101–1117. [CrossRef]
85. Frank, E.; Kramer, S. Ensembles of nested dichotomies for multi-class problems. In Proceedings of the Twenty-First International Conference on Machine Learning, Banff, AB, Canada, 4–8 July 2004; p. 39.
86. Rodríguez, J.J.; García-Osorio, C.; Maudes, J. Forests of nested dichotomies. *Pattern Recognit. Lett.* **2010**, *31*, 125–132. [CrossRef]
87. Quinlan, J.R. Bagging, boosting, and C4. 5. In Proceedings of the AAAI'96 Proceedings of the Thirteenth National Conference on Artificial Intelligence, Portland, OR, USA, 4–8 August 1996; Volume 1, pp. 725–730.
88. Rätsch, G.; Onoda, T.; Müller, K.R. Soft margins for AdaBoost. *Mach. Learn.* **2001**, *42*, 287–320. [CrossRef]
89. Breiman, L. Random forests. *Mach. Learn.* **2001**, *45*, 5–32. [CrossRef]
90. Rodriguez, J.J.; Kuncheva, L.I.; Alonso, C.J. Rotation forest: A new classifier ensemble method. *IEEE Trans. Pattern Anal. Mach. Intell.* **2006**, *28*, 1619–1630. [CrossRef] [PubMed]

91. Geurts, P.; Ernst, D.; Wehenkel, L. Extremely randomized trees. *Mach. Learn.* **2006**, *63*, 3–42. [CrossRef]

92. Cortes, C.; Vapnik, V. Support vector machine. *Mach. Learn.* **1995**, *20*, 273–297. [CrossRef]

93. Fauvel, M.; Tarabalka, Y.; Benediktsson, J.A.; Chanussot, J.; Tilton, J.C. Advances in spectral-spatial classification of hyperspectral images. *Proc. IEEE* **2013**, *101*, 652–675. [CrossRef]

94. Li, M.; Zang, S.; Zhang, B.; Li, S.; Wu, C. A review of remote sensing image classification techniques: The role of spatio-contextual information. *Eur. J. Remote Sens.* **2014**, *47*, 389–411. [CrossRef]

95. Chen, Y.; Zhao, X.; Jia, X. Spectral-spatial classification of hyperspectral data based on deep belief network. *IEEE J. Sel. Top. Appl. Earth Obs. Remote Sens.* **2015**, *8*, 2381–2392. [CrossRef]

96. He, L.; Li, J.; Liu, C.; Li, S. Recent advances on spectral-spatial hyperspectral image classification: An overview and new guidelines. *IEEE Trans. Geosci. Remote Sens.* **2018**, *56*, 1579–1597. [CrossRef]

97. Liao, W.; Chanussot, J.; Dalla Mura, M.; Huang, X.; Bellens, R.; Gautama, S.; Philips, W. Taking Optimal Advantage of Fine Spatial Resolution: Promoting partial image reconstruction for the morphological analysis of very-high-resolution images. *IEEE Geosci. Remote Sens. Mag.* **2017**, *5*, 8–28. [CrossRef]

98. Samat, A.; Persello, C.; Liu, S.; Li, E.; Miao, Z.; Abuduwaili, J. Classification of VHR Multispectral Images Using ExtraTrees and Maximally Stable Extremal Region-Guided Morphological Profile. *IEEE J. Sel. Top. Appl. Earth Obs. Remote Sens.* **2018**, *11*, 3179–3195. [CrossRef]

99. Samat, A.; Liu, S.; Persello, C.; Li, E.; Miao, Z.; Abuduwaili, J. Evaluation of ForestPA for VHR RS image classification using spectral and superpixel-guided morphological profiles. *Eur. J. Remote Sens.* **2019**, *52*, 107–121. [CrossRef]

100. Blaschke, T. Object based image analysis for remote sensing. *ISPRS J. Photogramm. Remote Sens.* **2010**, *65*, 2–16. [CrossRef]

101. Blaschke, T.; Hay, G.J.; Kelly, M.; Lang, S.; Hofmann, P.; Addink, E.; Queiroz Feitosa, R.; van der Meer, F.; van der Werff, H.; Tiede, D.; et al. Geographic object-based image analysis–towards a new paradigm. *ISPRS J. Photogramm. Remote Sens.* **2014**, *87*, 180–191. [CrossRef] [PubMed]

102. Ma, L.; Li, M.; Ma, X.; Cheng, L.; Du, P.; Liu, Y. A review of supervised object-based land-cover image classification. *ISPRS J. Photogramm. Remote Sens.* **2017**, *130*, 277–293. [CrossRef]

103. Kezer, K.; Matsuyama, H. Decrease of river runoff in the Lake Balkhash basin in Central Asia. *Hydrol. Process. Int. J.* **2006**, *20*, 1407–1423. [CrossRef]

104. Propastin, P.A. Simple model for monitoring Balkhash Lake water levels and Ili River discharges: Application of remote sensing. *Lakes Reserv. Res. Manag.* **2008**, *13*, 77–81. [CrossRef]

105. Propastin, P. Problems of water resources management in the drainage basin of Lake Balkhash with respect to political development. In *Climate Change and the Sustainable Use of Water Resources*; Springer: Berlin/Heidelberg, Germany, 2012; pp. 449–461.

106. Petr, T.; Mitrofanov, V.P. The impact on fish stocks of river regulation in Central Asia and Kazakhstan. *Lakes Reserv. Res. Manag.* **1998**, *3*, 143–164. [CrossRef]

107. Bai, J.; Chen, X.; Li, J.; Yang, L.; Fang, H. Changes in the area of inland lakes in arid regions of central Asia during the past 30 years. *Environ. Monit. Assess.* **2011**, *178*, 247–256. [CrossRef]

108. Klein, I.; Gessner, U.; Kuenzer, C. Regional land cover mapping and change detection in Central Asia using MODIS time-series. *Appl. Geogr.* **2012**, *35*, 219–234. [CrossRef]

109. Chen, X.; Bai, J.; Li, X.; Luo, G.; Li, J.; Li, B.L. Changes in land use/land cover and ecosystem services in Central Asia during 1990. *Curr. Opin. Environ. Sustain.* **2013**, *5*, 116–127. [CrossRef]

110. De Beurs, K.M.; Henebry, G.M.; Owsley, B.C.; Sokolik, I. Using multiple remote sensing perspectives to identify and attribute land surface dynamics in Central Asia 2001. *Remote Sens. Environ.* **2015**, *170*, 48–61. [CrossRef]

111. Zhang, C.; Lu, D.; Chen, X.; Zhang, Y.; Maisupova, B.; Tao, Y. The spatiotemporal patterns of vegetation coverage and biomass of the temperate deserts in Central Asia and their relationships with climate controls. *Remote Sens. Environ.* **2016**, *175*, 271–281. [CrossRef]

112. Leathart, T.; Pfahringer, B.; Frank, E. Building ensembles of adaptive nested dichotomies with random-pair selection. In *Joint European Conference on Machine Learning and Knowledge Discovery in Data Bases*; Springer: Cham, Switzerland, 2016; pp. 179–194.

113. Leathart, T.; Frank, E.; Pfahringer, B.; Holmes, G. Ensembles of Nested Dichotomies with Multiple Subset Evaluation. *arXiv* **2018**, arXiv:1809.02740.

114. Wever, M.; Mohr, F.; Hüllermeier, E. Ensembles of evolved nested dichotomies for classification. In Proceedings of the Genetic and Evolutionary Computation Conference, Kyoto, Japan, 15–19 July 2018; pp. 561–568.

115. Myint, S.W.; Gober, P.; Brazel, A.; Grossman-Clarke, S.; Weng, Q. Per-pixel vs. object-based classification of urban land cover extraction using high spatial resolution imagery. *Remote Sens. Environ.* **2011**, *115*, 1145–1161. [CrossRef]

116. Drăguţ, L.; Eisank, C. Automated object-based classification of topography from SRTM data. *Geomorphology* **2012**, *141*, 21–33. [CrossRef]

117. Drăguţ, L.; Csillik, O.; Eisank, C.; Tiede, D. Automated parameterisation for multi-scale image segmentation on multiple layers. *ISPRS J. Photogramm. Remote Sens.* **2014**, *88*, 119–127. [CrossRef]

118. Benz, U.C.; Hofmann, P.; Willhauck, G.; Lingenfelder, I.; Heynen, M. Multi-resolution, object-oriented fuzzy analysis of remote sensing data for GIS-ready information. *ISPRS J. Photogramm. Remote Sens.* **2004**, *58*, 239–258. [CrossRef]

119. Kim, M.; Warner, T.A.; Madden, M.; Atkinson, D.S. Multi-scale GEOBIA with very high spatial resolution digital aerial imagery: Scale, texture and image objects. *Int. J. Remote Sens.* **2011**, *32*, 2825–2850. [CrossRef]

120. Drăguţ, L.; Tiede, D.; Levick, S.R. ESP: A tool to estimate scale parameter for multiresolution image segmentation of remotely sensed data. *Int. J. Geogr. Inf. Sci.* **2010**, *24*, 859–871. [CrossRef]

121. Benediktsson, J.A.; Palmason, J.A.; Sveinsson, J.R. Classification of hyperspectral data from urban areas based on extended morphological profiles. *IEEE Trans. Geosci. Remote Sens.* **2005**, *43*, 480–491. [CrossRef]

122. Fauvel, M.; Benediktsson, J.A.; Chanussot, J.; Sveinsson, J.R. Spectral and spatial classification of hyperspectral data using SVMs and morphological profiles. *IEEE Trans. Geosci. Remote Sens.* **2008**, *46*, 3804–3814. [CrossRef]

123. Dalla Mura, M.; Villa, A.; Benediktsson, J.A.; Chanussot, J.; Bruzzone, L. Classification of hyperspectral images by using extended morphological attribute profiles and independent component analysis. *IEEE Geosci. Remote Sens. Lett.* **2011**, *8*, 542–546. [CrossRef]

124. Plaza, A.; Martinez, P.; Perez, R.; Plaza, J. A new approach to mixed pixel classification of hyperspectral imagery based on extended morphological profiles. *Pattern Recognit.* **2004**, *37*, 1097–1116. [CrossRef]

125. Aptoula, E.; Lefèvre, S. A comparative study on multivariate mathematical morphology. *Pattern Recognit.* **2007**, *40*, 2914–2929. [CrossRef]

126. Samat, A.; Gamba, P.; Liu, S.; Miao, Z.; Li, E.; Abuduwaili, J. Quad-PolSAR data classification using modified random forest algorithms to map halophytic plants in arid areas. *Int. J. Appl. Earth Obs. Geoinf.* **2018**, *73*, 503–521. [CrossRef]

127. Snoek, J.; Larochelle, H.; Adams, R.P. Practical bayesian optimization of machine learning algorithms. In Proceedings of the Advances in Neural Information Processing Systems, Lake Tahoe, NV, USA, 3–6 December 2012; pp. 2951–2959, Curran Associates.

128. Du, P.; Xia, J.; Zhang, W.; Tan, K.; Liu, Y.; Liu, S. Multiple classifier system for remote sensing image classification: A Review. *Sensors (Basel)* **2012**, *12*, 4764–4792. [CrossRef]

129. Samat, A.; Gamba, P.; Du, P.; Luo, J. Active extreme learning machines for quad-polarimetric SAR imagery classification. *Int. J. Appl. Earth Obs. Geoinf.* **2015**, *35*, 305–319. [CrossRef]

130. Vihervaara, P.; Auvinen, A.P.; Mononen, L.; Törmä, M.; Ahlroth, P.; Anttila, S.; Böttcher, K.; Forsius, M.; Heino, J.; Koskelainen, M.; et al. How essential biodiversity variables and remote sensing can help national biodiversity monitoring. *Glob. Ecol. Conserv.* **2017**, *10*, 43–59. [CrossRef]

131. Gholizadeh, H.; Gamon, J.A.; Zygielbaum, A.I.; Wang, R.; Schweiger, A.K.; Cavender-Bares, J. Remote sensing of biodiversity: Soil correction and data dimension reduction methods improve assessment of α-diversity (species richness) in prairie ecosystems. *Remote Sens. Environ.* **2018**, *206*, 240–253. [CrossRef]

132. Gholizadeh, H.; Gamon, J.A.; Townsend, P.A.; Zygielbaum, A.I.; Helzer, C.J.; Hmimina, G.Y.; Moore, R.M.; Schweiger, A.K.; Cavender-Bares, J. Detecting prairie biodiversity with airborne remote sensing. *Remote Sens. Environ.* **2019**, *221*, 38–49. [CrossRef]

Landslides Information Extraction Using Object-Oriented Image Analysis Paradigm Based on Deep Learning and Transfer Learning

Heng Lu [1,2], **Lei Ma** [3,4,5], **Xiao Fu** [6], **Chao Liu** [1,2], **Zhi Wang** [1,2], **Min Tang** [7] and **Naiwen Li** [1,2,*]

[1] State Key Laboratory of Hydraulics and Mountain River Engineering, Sichuan University, Chengdu 610065, China; luheng@scu.edu.cn (H.L.); liuchao@scu.edu.cn (C.L.); zhizhsally@163.com (Z.W.)
[2] College of Hydraulic and Hydroelectric Engineering, Sichuan University, Chengdu 610065, China
[3] School of Geography and Ocean Science, Nanjing University, Nanjing 210093, China; maleinju@nju.edu.cn
[4] Key Laboratory for Satellite Mapping Technology and Applications of State Administration of Surveying, Mapping and Geoinformation, Nanjing University, Nanjing 210023, China
[5] Signal Processing in Earth Observation, Technical University of Munich (TUM), 80333 Munich, Germany
[6] Faculty of Geosciences and Environmental Engineering, Southwest Jiaotong University, Chengdu 611756, China; fuxiao@my.swjtu.edu.cn
[7] China Railway Eryuan Engineering Group Co., Ltd., Chengdu 610031, China; tangmin05@ey.crec.cn
[*] Correspondence: linaiwen@scu.edu.cn; Tel.: +86-138-8222-1763

Received: 31 December 2019; Accepted: 22 February 2020; Published: 25 February 2020

Abstract: How to acquire landslide disaster information quickly and accurately has become the focus and difficulty of disaster prevention and relief by remote sensing. Landslide disasters are generally featured by sudden occurrence, proposing high demand for emergency data acquisition. The low-altitude Unmanned Aerial Vehicle (UAV) remote sensing technology is widely applied to acquire landslide disaster data, due to its convenience, high efficiency, and ability to fly at low altitude under cloud. However, the spectrum information of UAV images is generally deficient and manual interpretation is difficult for meeting the need of quick acquisition of emergency data. Based on this, UAV images of high-occurrence areas of landslide disaster in Wenchuan County and Baoxing County in Sichuan Province, China were selected for research in the paper. Firstly, the acquired UAV images were pre-processed to generate orthoimages. Subsequently, multi-resolution segmentation was carried out to obtain image objects, and the barycenter of each object was calculated to generate a landslide sample database (including positive and negative samples) for deep learning. Next, four landslide feature models of deep learning and transfer learning, namely Histograms of Oriented Gradients (HOG), Bag of Visual Word (BOVW), Convolutional Neural Network (CNN), and Transfer Learning (TL) were compared, and it was found that the TL model possesses the best feature extraction effect, so a landslide extraction method based on the TL model and object-oriented image analysis (TLOEL) was proposed; finally, the TLOEL method was compared with the object-oriented nearest neighbor classification (NNC) method. The research results show that the accuracy of the TLOEL method is higher than the NNC method, which can not only achieve the edge extraction of large landslides, but also detect and extract middle and small landslides accurately that are scatteredly distributed.

Keywords: landslides information extraction; unmanned aerial vehicle imagery; convolutional neural network; transfer learning; object-oriented image analysis

---

1. Introduction

Human beings are facing, and will continue to face, challenges that affect the harmony and sustainable development of the society for a long time, such as population, resources, and environment.

Among all environmental problems, geological environment is one of the most prominent [1–3]. For one thing, the geological environment is a necessary carrier and a basic environment for all human life and engineering activities. For another, it is fragile and difficult, or even impossible to restore. Landslide is a dire threat to people's lives and property and social public safety [4–6]. Geological disasters very frequently occur in China and cause tremendous loss, especially in the western mountainous areas with complex topographic and geological conditions. Landslides in these areas are generally characterized by suddenness. There is no forewarning that can be directly observed and perceived beforehand since landslide is triggered by external factors (such as heavy rainfall, earthquakes, etc.). Therefore, quick and automatic information extraction of sudden landslides has become a hot topic, also a hot potato, of the day in the landslide research in the world [7,8].

In the field of remote sensing, the extraction of landslide information that is based on satellite images or aerial images is mainly realized through the spectrum, shape, and texture features of landslides that are shown in images that are different from other surface features [9]. In the early application, the identification of landslide information and boundary extraction were mainly actualized by means of manual digitization. This method is featured by high accuracy, but, when it is necessary to process the data of a large region or to meet the disaster emergency demand, the manual digitalization operation mode is of no advantage in time and cost. In addition, if the region is segmented into several regions for different interpreters to interpret, it is inevitable that the subjectivity of different interpreters will be brought into the interpretation results [10,11]. With the development of digital image processing technology, increasing image classification algorithms have been applied to the extraction of landslide information. The reflectivity difference (spectrum information) of different surface features on the remote sensing image is used to extract the landslide region and non-landslide region. Generally, the landslide region shows a high reflectivity on the remote sensing image, which is easy to be distinguished from surface features with low reflectivity, but it is easy to be confused with bare land because the bare land also has high reflectivity [12]. Additionally, this kind of pixel-oriented method makes the classification result easily produce "salt and pepper" noise. In recent years, with the launch of more and more earth observation satellites with high-resolution sensors, the data sources for research are more and more abundant, and the object-oriented image analysis method comes into being. In the object-oriented image analysis method, the available information, such as spectrum, shape, texture, context semantics, and terrain on remote sensing images are comprehensively selected to extract the information of surface features [13–17].

In recent years, with the development of machine learning technology, more and more algorithms have been applied to the remote sensing identification of landslides. At present, the widely used machine learning algorithms include support vector machine (SVM), random forest (RF), artificial neural network (ANN), convolutional neural network (CNN), deep convolutional neural network (DCNN), etc. [18–23]. The conventional object-oriented image analysis method requires acquiring a large number of image features for subsequent classification, and carrying out a large number of feature selection experiments, which is very time-consuming and difficult to obtain accurate features completely. Deep learning (DL) and transfer learning (TL) are the fastest-developing machine learning methods that are applied to remote sensing image classification in recent years, which can automatically extract features from the original images and the extracted deep features are often very effective for processing complex images [24–26]. However, the problem is that the features outputted by the deep learning method are highly abstract, and the boundaries of actual surface features cannot be accurately obtained, and the classification result is different from the boundaries of the actual surface features. However, the object-oriented image analysis method is based on homogeneity to segment and obtain the boundaries of surface features, usually the results of segmentation are consistent with the actual boundaries of surface features. Therefore, a method integrating transfer learning and object-oriented image analysis is proposed in this paper by combining the respective advantages of the two methods. Firstly, the multi-resolution segmentation algorithm obtains an irregular segmented object, and then a regular image block is generated according to the barycenter of the segmented object, so that the segmented object that is obtained by the object-oriented method is combined with the transfer learning. The rest of this paper are organized, as follows: In Section 2, three research sites selected in the experiment are introduced. In Section 3, the availability of four deep learning models Histograms of Oriented Gradients (HOG), Bag of Visual Word (BOVW), CNN, and TL) for landslide feature extraction were compared, and the realization process of the proposed method and the experiment steps are described in detail. The experiment results are given in Section 4 and discussed in Section 5. Finally, the full text is summarized in Section 6.

2. Study Sites

The experimental area is located in Wenchuan and Baoxing counties in Sichuan province, China. The experimental data contains high-risk areas of geological hazards UAV images. The images with spatial resolution of 0.2 m were taken in May 2017. The UAV that is used in the experiment is a fixed wing, which is equipped with SONY Cyber shot DSC RX1 digital camera. The relative flying height predetermined for this experiment is 600 m, the longitudinal overlap is 70%, and the lateral overlap is 40%. Small squares (marked in green, blue, and red) are selected for the experimental purpose, as shown in Figure 1.

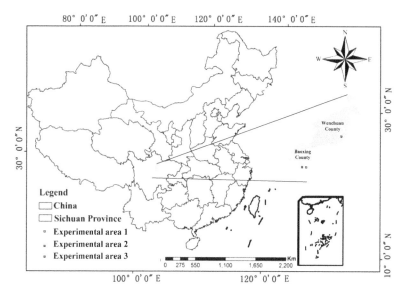

Figure 1. Location of study area.

3. Methods

Firstly, the acquired UAV images were pre-processed to generate orthoimages. Subsequently, the image objects were obtained by multi- resolution segmentation, and calculating the barycenter of each object was undertaken to generate the samples for deep learning landslide (including positive and negative samples). Next, HOG, BOVW, CNN, and TL landslide feature models were compared, and found that the TL model had the best feature extraction effect. Therefore, TL model and object-oriented image analysis (TLOEL) method based on TL model and object-oriented image analysis was proposed. Finally, the TLOEL method was compared with the NNC method. Figure 2 shows the research process of this paper.

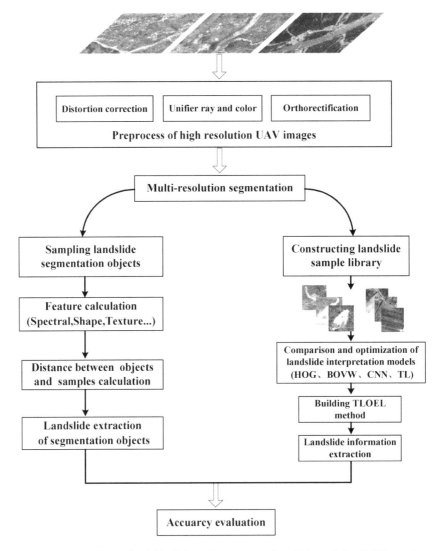

Figure 2. The workflow of landslides information extraction from high-resolution UAV imageries.

3.1. Preprocess of High-Resolution Images

The digital camera on UAV is of non-metric type, so the images are subject to serious lens distortion. Therefore, distortion correction shall be carried out based on distortion parameters of the camera. Meanwhile, the exposure time intervals and different weather conditions in the flight course will result in chromatic aberration, so color and light uniformizing shall be carried out with the mask method. Preliminary image sorting and positioning can be carried out for matching homologous points of adjacent image pairs based on the aircraft attitude parameters recorded by the flight control system. After the matching of homologous points, block adjustment can be made based on the conditions of collinearity equation. After that, the coordinates of ground control points may be incorporated to realize absolute orientation, so as to obtain the corrected orthoimages [27,28].

3.2. Segmentation

Image segmentation is the first step of the experiment to form a basic classification unit (object) with high homogeneity. Multi-resolution segmentation has proved to be a successful segmentation algorithm in many applications. In the research stated in this paper, the multi-resolution segmentation algorithm of uniform rule is used in the eCognation 9.0 software to generate the segmented objects for the three experiment areas [29]. Using the multi-resolution segmentation algorithm requires setting three parameters: segmentation scale, color/shape ratio, and smoothness/compactness ratio. Among them, the most important parameter is segmentation scale, which determines the heterogeneity inside the object. Specifically, when the segmentation scale is too large to the classification target object, undersegmentation occurs, and small objects are "submerged" by large objects, thus resulting in mixed objects. When the segmentation scale is too small to the classification target object, over-segmentation occurs, which causes the segmentation result to be "broken" and increases the calculation burden of the subsequent classification process. The color/shape ratio reflects the ratio of spectral uniformity to shape uniformity. The smoothness/compactness ratio is used to define the smoothness or compactness of each object.

3.3. Constructing Landslide Sample Library

Image objects with irregular boundaries must be transformed into image blocks with regular shapes and fixed sizes in order to combine the TL model with the object-oriented image analysis method. The size of image block is related to the depth of CNN and it is limited to computer hardware (e.g., memory capacity). Through experiments and comparative literature, it is found that it is most suitable to choose 256×256 pixels as the size of image block instead of the super large CNN learning frame [30].

During the experiment, it is found that there is a lot of work to manually build a landslide sample library with positive and negative samples, so this paper studies how to realize automatic batch cutting of image blocks that are based on ArcGIS Python secondary development package (ArcPy). The image blocks after cutting are divided into positive and negative samples by visual interpretation.

The specific process flow for building a landslide sample library is as follows:

(1) Calculate a barycentric point position of each segemented objects, and take the barycentric point position as the center of the image block.

(2) Automatically generate the boundary of image block on the ArcPy platform, and clip to generate an image block according to the boundary and store it.

(3) Visually identify and distinguish an image block containing a landslide, storing the image block as a positive sample, and storing the remaining image blocks as negative samples, and removing the image blocks with no practical significance or the image blocks with too cluttered surface features.

3.4. Building Landslide Interpretation Model

3.4.1. Landslides Feature Extraction Based on HOG Model

Dalal proposed HOG on CVPR in 2005 [31]. When compared with deep learning, it is a common shallow feature that is used in computer vision and pattern recognition to describe the local texture of an image, applied as a feature descriptor to perform object detection. Figure 3 shows the process of HOG feature extraction.

First, divide the images into several blocks by statistics. Afterwards, calculate the distribution of edge intensity histogram separately. Finally, assemble the block histograms together to get feature descriptors of the images. As HOG operates on the local grid unit of the image and the space field is small, it can keep well the original image geometry and optical deformation. The combination of HOG features and SVM classifiers has been widely used in image identification. The HOG feature can be used to distinguish them while considering that there is a gradient change between the landslide area and the surrounding environment in a high-resolution image.

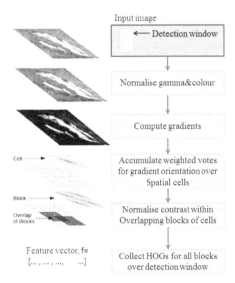

Figure 3. The flow chart of Histograms of Oriented Gradients (HOG) feature extraction.

3.4.2. Landslides Feature Extraction Based on BOVW Model

BOVW is an image representation model using Bag of Words (BOW). It can map two-dimensional image information into a set of visual keywords, which effectively compresses the description of the image and saves the local features of the image. The BOVW model extracts low-level features from images in the sample library, and then, given the number of cluster centers, clusters these low-level features with an unsupervised algorithm, such as K-means [32]. Here is a sequence of observations $(x_1, x_2, \ldots, x_n)$. Each of the observed value is a d-dimensional real-value vector. The goal of K-means is to divide these n observations into k sequences $s = (s_1, s_2, \ldots, s_k)$, $k < n$, such as:

$$\text{argmin} \sum_{i=1}^{k} \sum_{x_j \in S_i} \|x_j - \mu_i\| \tag{1}$$

where μ_i is the average of s_i.

Visual keywords ("vocabulary: represented by w1, w2, ..., w_p, w_r, ..., w_m" in Figure 5) are obtained through the cluster centers, mapping each feature extracted from the images to the nearest visual vocabulary, where the image can be represented as a histogram feature descriptor. Figure 4 shows the feature extraction process.

3.4.3. Landslides Feature Extraction Based on CNN Model

CNN was mainly used to identify distortion-invariant (e.g. displacement, scaling) two-dimensional graphics in previous applications [33]. As CNN's feature detection layer learns with training data, implicit feature extraction is adopted instead of the explicit one when CNN is used. Neurons of the same feature map are set to have the same weights, so the network can learn in parallel, which is a big advantage of CNN relative to a simple neural network [34,35].

Feature extraction is to abstract the image information and obtain a set of feature vectors that can describe the image. Extraction is the key to image classification and selection determines the final result of classification. Manual visual interpretation can produce good classification results for new data and tasks, accompanied by large workload, low efficiency, subjective randomness, and non-quantitative analysis [36]. The middle and low-level features perform well in specific classification and identification tasks. However, high-resolution image is a far cry from an ordinary natural image—the spatial spectrum changes greatly, so the middle and low levels feature extraction is not able to produce a good result in high-resolution images. With the continuous advancement in deep learning, by inputting data to extract the features layer-by-layer from bottom to top, mapping the relationships between bottom signals and top lexemes can be established, so that the high-level features of the landslide can be obtained to better present the landslide in a high-resolution image [37].

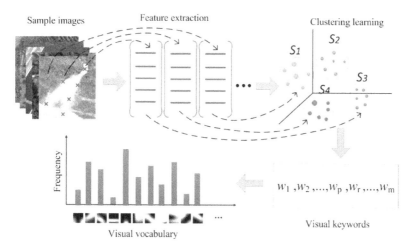

Figure 4. The flow chart of Bag of Visual Word (BOVW) feature extraction.

CNN avoids explicit feature sampling and learns implicitly from the training data, which distinguishes it from other neural-network-based classifiers. Feature extraction is integrated into the multi-layer perceptron through structural reorganization and weights reduction, so that gray-scale images can be directly processed and, therefore, CNN can be directly used to process image-based classification [38]. It has many advantages in image processing. (1) Input images match network topology perfectly. (2) Feature extraction and pattern classification are performed simultaneously and simultaneously generate results in training. (3) Weight sharing reduces the training parameters of the network and makes the structure of the neural network simpler and more adaptable [39,40]. A basic convolutional neural network structure can be divided into three layers, namely feature extraction layer, feature mapping layer, and feature pooling layer. A deep convolutional network can be established by stacking multiple basic network structures, as shown in Figure 5. Conv1 and conv2 represent convolutional layer 1 and convolutional layer 2, and pool1 and pool2 represent pooling layer 1 and pooling layer 2.

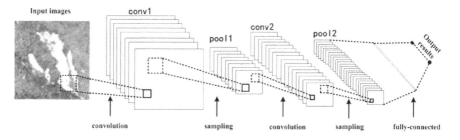

Figure 5. Construction of deep convolution neural network.

(1) Feature extraction layer: This layer is the feature extraction layer. The input of each neuron is connected with the local receptive field of the previous layer and extracts the local features. Assume that the input image I is a two-dimensional matrix, the size of which is $\gamma \times c$, use a trainable filter group k, the size of which is $w \times w$, to compute the convolution, and the step size is l. Finally, there is a Y output of $((\gamma - w)/l + 1) \times ((c - w)/l + 1)$ size. Where:

$$y_i = b_i + \sum_i k_{ij} * x_i \tag{2}$$

where x_i represents the input convolutional layer, k_{ij} represents the convolutional kernel parameters, b_i represents the deviation value, and * represents the convolution operation. Each filter corresponds to a specific feature.

(2) Feature mapping layer: A nonlinear function is used to map the results of the filter layer to ensure the validity of the feature, and the feature map F is obtained.

$$f_i = \delta(b_i + \sum_i k_{ij} * x_i) \tag{3}$$

where, δ is the activation function. Tanh, sigmoid, and softplus are common activation functions. Tanh is a variant of sigmoid, whose value range is [0,1]. The linear correction unit ReLU is the closest to the activation model of biological neurons after stimulation and has certain sparsity. The calculation is simple, which is helpful to improve the effect [41].

(3) Feature pooling layer: Theoretically, features can be acquired through convolution and then directly used to train the classifier. However, the feature dimensions of any medium-sized image are in millions after convolution, and the classifier is easily overfitted after direct training. Therefore, the pooling of convolution features, or downsampling, is needed. F is the convolution feature map, which is divided into disjoint regions with a size of $m \times m$, and then calculate the average value (or maximum value) of these regions and taken as the pooling feature P, whose size is $\{((\gamma - w)/l + 1)/m\} \times \{((c - w)/l + 1)/m\}$. The pooled feature dimension is greatly reduced to avoid overfitting and it is robust. Figure 6 is the flow chart of the CNN-based landslide interpretation model.

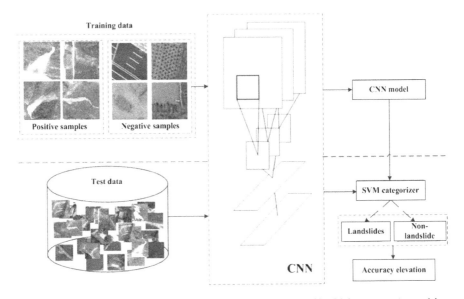

Figure 6. The flow chart of the convolutional neural network (CNN)-based landslide interpretation model.

3.4.4. Landslides Feature Extraction Based on TL Model

In the process of training a deep learning model, problems, such as insufficient training samples, often occur. The emergence of TL has aroused extensive attention and research. Technologies that are related to machine learning and data mining have been used in many practical applications [42]. In traditional machine learning framework, the task of learning is to learn a classification model with sufficient training data. In the field of image identification, the first step is to label a large amount of image data manually and, then, based on the machine learning method, obtain a classification model, which is used to classify and identify the test images. Traditional machine learning needs to calibrate a lot of training data for each field, which will cost a lot of manpower and material resources. However, without a large number of labeled data, many studies and applications that are related to learning cannot be carried out. Generally, traditional machine learning assumes that training data and test data obey the same data distribution. However, in many cases, this assumption cannot be met, which might lead to the expiration of training data. This often requires re-labeling a large number of training data to meet the needs of training, but it is very expensive to label new data because it requires a lot of manpower and material resources.

There are many connections between the target detection of remote sensing image and natural image in nature; in many ways, they are thought of as the same problem. The goal of transfer learning is to transfer knowledge from the existing priori sample data and use the knowledge learned from an environment to help the learning task in the new environment. Moreover, there is no strict assumption, as traditional machine learning theory requires that the training data and test data should have the same distribution [43]. The weight of a new category and a classification model applicable to the target task are obtained through pre-training models in the existing classification data set, removing the neural network on top of the training model, and retraining an output layer through the target task data set. This method can shorten the training time of the model and improve work efficiency [44,45]. At present, there are many labeled natural image libraries. For example, the typical ImageNet library labeled by Stanford University, which contains millions of labeled images, includes more than 15 million high-resolution images with labels. These images are divided into more than 22,000 categories, thus making it the largest labeled image library in image identification field. The pre-training model is obtained by learning the method of image feature extraction from ImageNet library with the method of transfer learning on features.

Figure 7 is the framework of the landslide interpretation model that is described in this paper, which is obtained through transfer learning. It mainly includes three parts, namely feature learning, feature transfer, and landslide interpretation model training. Source task is the scene classification in the original deep learning. The classification model of the target task is built by transferring the network parameters and the results of the source task to the optimized target task. In Figure 7, conv represents convolutional layer, pool represents pooling layer, and FC represents the fully-connected layer.

3.4.5. Reliability Evaluation of Landslide Feature Extraction Model

There are many methods for evaluating the landslide feature extraction results. The Confusion Matrix is used to verify the accuracy of the interpretation model, in accordance with the quantitative research needs of this paper [46]. Table 1 shows the evaluation indicator system.

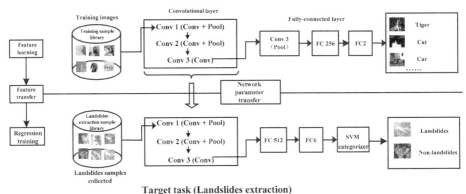

Figure 7. Feature extraction of landslides by transfer learning.

Table 1. Related indicators of confusion matrix.

		Prediction		Total
		1	0	
Actual	1	True Positive (TP)	False Negative (FN)	Actual Positive Prediction (TP+FN)
	0	False Positive (FP)	True Negative (TN)	Actual Negative Prediction (FP+TN)
Total		Positive Prediction (TP+FP)	Negative Prediction (FN+TN)	Total (TP+FP+FN+TN)

Where TP represents the positive is predicted to be a positive value; TN indicates that the negative is predicted to be a negative value; FP means that the negative is predicted to be a positive value; and, FN indicates that the positive is predicted to be a negative value. Precision, ACC, and Recall rate are defined, as shown in Formula (4).

$$
\begin{aligned}
FPR &= \frac{FP}{FP + TN}; \\
TPR &= \frac{TP}{TP + TN}; \\
ACC &= \frac{TP + TN}{Total}.
\end{aligned}
\tag{4}
$$

3.5. Combination of Object-oriented Image Analysis and TL Model

The core concept of the object-oriented classification method is that the single pixel is taken as the minimum cell under a specific scale and calculated with its neighborhood pixels in the principle of being the most suitable for each other to obtain the object of segmentation with the best homogeneity; when the segmentation on a certain scale has been completed, the new object of segmentation is taken as a cell to continue the calculation of adjacent objects and then merge to generate the object under the new scale until merging under such scale has been completed [47]. As the object-oriented classification method regards the body as an integral object, such body also has such features as spatial form, geometric length, and neighborhood relationship, in addition to the spectral characteristic under this ideological system and, therefore, its accuracy is lifted to a certain extent in comparison with that of the pixel-oriented method. Although making use of object-oriented image segmentation technology can effectively realize segmentation in ground object target, it is difficult to obtain the specific attribute information on the ground object. The structure of landslide has certain space geometry and texture features, yet it is very difficult to accurately identify the landslide from mass and abundant remote sensing data by the traditional classification method and it possesses obvious advantages to realize the structural features of a complex system by CNN interpretation.

In the process of landslides information extraction from high-resolution images, the feature extraction determines the final accuracy. A great number of samples are required to support the feature extraction due to the huge difference between high-resolution images and ordinary natural images and great change in spatial spectrum, but there is a limited number of landslides in the research region and less collectible samples and, hence, this article combines the aforesaid TL model and the object-oriented image analysis to establish a method for landslide information extraction from high-resolution UAV images that can realize large-scale scattered landslide information extraction. The process of TLOEL method is as shown in Figure 8.

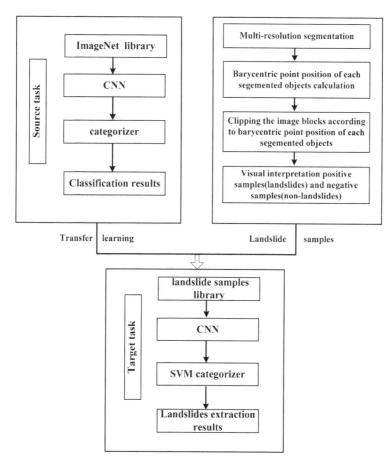

Figure 8. The flow chart of landslides extraction by the transfer learning model and object-oriented image analysis (TLOEL) method.

3.6. Landslide Information Extraction by NNC Method

The NNC method is a mature classification method in view of its simple operation, high efficiency, and wide application scope [48]. The control group researched in the paper is the nearest neighbor classification method. The specific process is as follows: First, the sample object is selected and statistically analyzed to obtain relevant feature values, such as texture, spectrum, shape, and neighborhood information, so as to build a multi-dimensional feature space. Afterwards, the distance difference between the object to be classified and the sample is calculated, and the object to be classified is classified into the class according to the distance relationship of the features and the membership function to determine which sample class is nearest to the object to be classified. As shown in Formula (5).

$$d = \sqrt{\sum_f \left[\frac{v_f^{(s)} - v_f^{(\sigma)}}{\sigma_f} \right]^2} \tag{5}$$

Where, d is feature distance, f is the feature, s is the sample object, σ is the object to be classified, σ_f is the standard deviation of feature f value, $v_f^{(s)}$ is the feature value of feature f of sample object s, and $v_f^{(\sigma)}$ is the feature value of feature f of object to be classified σ.

3.7. Accuarcy Evaluation

The accuracy of information extraction from different ground objects is known as the classification accuracy, and it is a common standard for testing the degree of advantages and disadvantages of the classification rules. We usually modify the final classification results according to the result of assessment on the classification accuracy, and if such assessment is low, it is necessary to improve the rule definition. The methods for assessment on the classification accuracy generally fall into two categories: one refers to the qualitative accuracy assessment methods by artificial visual interpretation and the other refers to the quantitative accuracy assessment methods [49]. The artificial visual interpretation method gives consideration to certain reliability on the premise of rapid assessment, but only interpretation professionals can carry out related operation, which results in large subjectivity of the assessment results; serious divergence between the results evaluated by field investigation and visual interpretation, respectively, indicates an undesirable classification, so it is required to set the feature rules for reclassification and the accuracy assessment will not be made until the classification results are relatively identical. The accuracy assessment is usually shown in percentage and the accuracy assessment method widely applied at present is called the confusion matrix method, which is defined, as follows:

$$M = \begin{bmatrix} m_{11} & m_{12} & \cdots & m_{1n} \\ m_{21} & m_{22} & \cdots & m_{2n} \\ \vdots & & \ddots & \vdots \\ m_{n1} & m_{n2} & \cdots & m_{nn} \end{bmatrix} \tag{6}$$

where m_{ij} represents the total number of pixels which are assigned to Category j from those subordinate to Category i in the research region and n represents the total number of categories. In the confusion matrix, the greater value in the leading diagonal indicates a higher reliability in the classification results.

The common indexes of assessment on classification accuracy include overall accuracy, product's accuracy, user's accuracy, and Kappa coefficient in addition to the confusion matrix.

(1) Overall Accuracy (OA) refers to the specific value of a total number of all correct classifications and that of samplings and reflects the degree of correctness of all categories in the classification results of images. It is calculated in the following formula:

$$OA = \frac{\sum_{i=1}^{n} m_{ii}}{\sum_{j=1}^{n} \sum_{i=1}^{n} m_{ij}} \tag{7}$$

(2) Product's Accuracy (PA) refers to the specific value of the number of pixels in correct classification from a single category and the total number of pixels in reference data of such category. It is calculated in the following formula:

$$PA = \frac{m_{ii}}{\sum\limits_{j=1}^{n} m_{ij}} \tag{8}$$

(3) User's Accuracy (UA) refers to the specific value of the number of pixels in correct classification from a single category and the total number of pixels in such category and it indicates the probability that a classified pixel authentically represents such category. It is calculated in the following formula:

$$UA = \frac{m_{ii}}{\sum\limits_{j=1}^{n} m_{ji}} \tag{9}$$

(4) The Kappa coefficient refers to an assessment index to judge the extent of coincidence between two images and range from 0 to 1. It indicates how much the classification method selected is better than the method that the single pixel is randomly assigned to any category. It is calculated in the following formula:

$$K = \frac{N \sum\limits_{i=1}^{n} m_{ii} - \sum\limits_{i=1}^{n} m_{i+}m_{+i}}{N^2 - \sum\limits_{i=1}^{n} m_{i+}m_{+i}} \tag{10}$$

where n represents the total number of categories, m_{ij} represents the pixel value at Line i and Row j in the confusion matrix, N represents the total number of samples, and m_{+i} and m_{i+} are, respectively, sums of rows and lines in the confusion matrix. The Kappa coefficient is calculated by comprehensive utilization of all information in the confusion matrix and, therefore, can be used as a comprehensive index for an assessment on classification accuracy.

4. Results

4.1. Preprocessing of High-Resolution Images

First, make a correction of the image distortion based on the camera's distortion parameters, and then conduct uniform color and light processing by mask method and carry out preliminary sequencing and locating for homonymy points matching of adjacent image pair by virtue of aircraft attitude parameter data recorded in the flight control system. Finally, make the block adjustment according to the collinearity equation condition. After the completion of the block adjustment, add the coordinates of ground control points to achieve absolute orientation and then obtain the corrected orthoimage. Figure 9 shows the preprocessed UAV images.

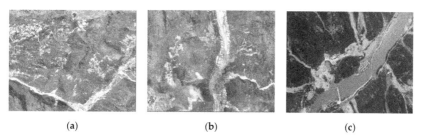

| (a) | (b) | (c) |

Figure 9. The preprocessed UAV imageries. (**a**) Experimental image 1; (**b**) Experimental image 2; and, (**c**) Experimental image 3.

4.2. Segmentation Results

This paper uses eCognition 9.0 software for multi-resolution segmentation of the experimental regions and an ESP tool determined the optimal dimension in the process of segmentation. Table 2 shows the optimal image segmentation parameters and Figure 10 shows the segmentation results.

Table 2. Optimal image segmentation parameters.

Experimental Images	Segmentation Scale	Color/Shape	Smoothness/Compactness	Number of Image Objects
Experimental image 1	50	0.4/0.6	0.5/0.5	490
Experimental image 2	40	0.4/0.6	0.5/0.5	434
Experimental image 3	60	0.4/0.6	0.5/0.5	330

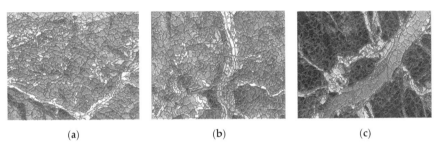

(a)   (b)   (c)

Figure 10. Segmentation results. (**a**) Segmentation result of experimental image 1; (**b**) Segmentation result of experimental image 2; and, (c) Segmentation result of experimental image 3.

4.3. Establishment of Landslide Sample Library

Based on the principle that is mentioned in Section 3.3, establish the batch of items by ArcPy for batch processing of the images from the research region. Take the barycentric point position as the center of the image block to automate clipping, numbering, and storage. Figure 11 shows the partial of sample examples clipped and stored.

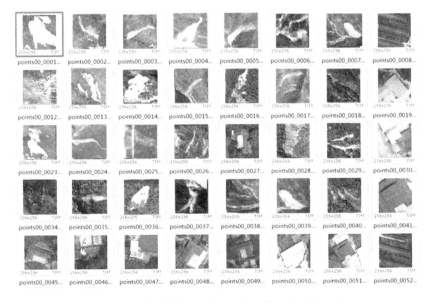

Figure 11. Partial of sample examples clipped and stored.

It should be illustrated that the landslide samples are naturalized to a size of 256*256 in the experiment for the convenience of rapid computer processing (feature calculation) by giving consideration to both the computer processing ability in the identification process and the spatial resolution of landslide sample images. This article makes the certain screening of the samples collected by the visual interpretation method. A large number of positive and negative samples were marked for different types of landslide in the experimental region, and the samples were expanded by means of rotation and mirroring. With the increase in the number of positive samples of landslides in different types and structures, the deep learning algorithm can express the landslide features in a deeper level, and then obtain the landslide information in a more accurate and effective manner from the image, thus providing higher accuracy of recognition. Finally, the sample library established includes 5000 positive and 10000 negative samples of landslides, which are all zoomed to a pixel size of 256×256. Figure 12 shows the sample examples obtained by the above method.

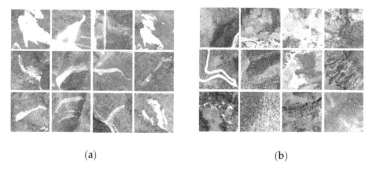

(a)   (b)

Figure 12. Example sample library. (**a**) Positive samples (**b**) Negative samples.

4.4. Results of Landslide Interpretation Model Construction

The HOG parameters in the experiment are as follows: Bin with eight histograms are used for the gradient histogram projection, a single cell is in a size of 32×32 pixels, each Block is composed of 2×2 Cells, the sample size is unified as 256×256 pixels, and the length of feature vector finally obtained by merging is 1568. Figure 13 shows the visualization results of the HOG feature of experimental landslide samples.

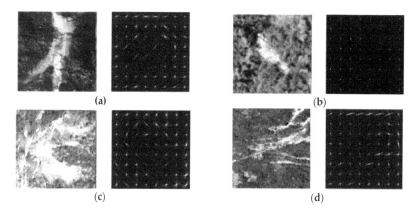

(a)   (b)

(c)   (d)

Figure 13. The visualization results of HOG feature of experimental landslide samples. (**a**) landslide sample 1 and HOG feature; (**b**) landslide sample 2 and HOG feature; (**c**) landslide sample 3 and HOG feature; (**d**) landslide sample 3 and HOG feature.

BOVW parameters in the experiment are as follows: 128-dimension Surf feature is extracted from local features in the experiment, the "dictionary" is established based on K-means in the process of dictionary learning and the number of dictionaries is set as 400 [50]. Finally, find the Surf feature in the dictionary to generate statistics of its histogram distribution as the BOVW feature of the image.

This article uses the ImageNet classification model Decaf pre-trained by the computer vision research group of the University of California Berkeley in establishment of TL feature at the stage of feature learning and it is composed of three convolutional layers and three fully-connected layers. At the stage of feature transfer, the output of fc6 layer is selected as the feature on the conditions that the parameters of three convolutional layers and the fc6 layer of the whole model are kept constant, as the research results of such group indicate that the output of fc6 layer has a better generalization ability when compared with that of other layers and the feature vector is in a proper length (4096). The feature vector that is obtained based on transfer learning serves as the training input of SVM and it is used for training of landslide information extraction model. This article establishes a landslide interpretation model that is based on feature transfer and makes a feasibility analysis of whether the transferred feature extraction method is suitable for landslide interpretation in consideration of the landslide data of the research region. The experiment uses t-SNE (t-Distributed Stochastic Neighbor Embedding) method to achieve the visualization of sample clustering of TL landslide interpretation model; as a data dimension reduction method, t-SNE can offer a visual representation of data point positions of high dimensional data in two-dimensional or three-dimensional space, similar to PCA, and it is widely applied for high dimensional visualization in the fields of computer vision and machine learning.

Figure 14 shows the t-SNE visualization result of the landslide interpretation model feature based on transfer learning feature. Here the feature vector in 4096 dimensions is mapped into two-dimensional space by the t-SNE method to obtain the distribution model of the transfer learning feature. It can be seen from the red circle region in Figure 14 that, in the transfer learning feature space, the landslide samples have presented obvious aggregation, and it is believed that the feature extraction method that is obtained from training and learning of original image library has explored the mapping relation between natural and landslide images, which indicates that it is feasible to establish the landslide interpretation model based on transfer learning.

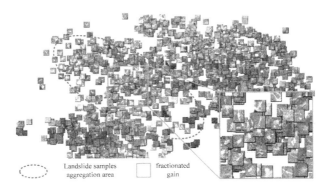

Figure 14. The t-Distributed Stochastic Neighbor Embedding (t-SNE) visualization result of landslide interpretation model feature based on TL feature.

The experiment uses the SVM as the output classifier and it takes the square root error L2 as a loss function by use of linear kernel function and L2 regularization. 10% samples (500 positive and 1000 negative samples) are left for testing in the process of training. The training results are as follows: the optimal parameters for cross validation of HOG, BOVW, CNN, and TL features are, respectively, {'C': 8000}, {'C': 6000}, {'C': 3000}, and {'C': 4000}, and the key parameter C is obtained by six-fold (six folders) cross validation. Table 3 shows the confusion matrix.

Table 3. Confusion matrix.

Parameters	HOG		BOVW		CNN		TL	
	Landslides	Non-landslides	Landslides	Non-landslides	Landslides	Non-landslides	Landslides	Non-landslides
Landslides	376	124	431	69	489	11	492	12
Non-landslides	189	811	117	883	25	975	18	978
Precition/%	66.5		78.6		95.1		96.4	
Recall rate/%	75.2		86.2		97.8		97.6	
ACC/%	79.1		87.6		97.6		98	

It can be seen from the training results in Table 2 that CNN (Precision: 95.1%, ACC: 97.6%) and TL (Precision: 96.4%, ACC: 98%) have certain advantages in high-resolution images when compared with BOVW (Precision: 78.6%, ACC: 87.6%) and HOG (Precision: 66.5%, ACC: 79.1%), so this article selects CNN and TL as the extraction features to establish the landslide extraction model.

4.5. Landslides Information Extraction

Based on the principle that is mentioned in Sections 3.5 and 3.6, Figure 15 shows the results of NNC method and the TLOEL method in the three experimental areas.

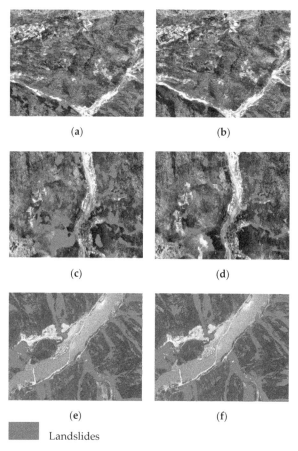

(a)   (b)

(c)   (d)

(e)   (f)

Landslides

Figure 15. Results of landslide information extraction based on two methods. (**a**) Experimental image 1, based on NNC; (**b**) Experimental image 1, based on TLOEL; (**c**) Experimental image 2, based on NNC; (**d**) Experimental image 2, based on TLOEL; (**e**) Experimental image 3, based on NNC; and, (**f**) Experimental image 3, based on TLOEL.

While considering the high spatial resolution of UAV images, the validation data can be directly obtained by visual interpretation combined with field investigation. For experimental image 1, 280 landslide data verification points and 320 non-landslide data verification points were randomly acquired by manual visual interpretation and field investigation. 320 landslide data verification points and 350 non-landslide data verification points were randomly acquired by manual visual interpretation and field investigation for experimental image 2. For experimental image 3, 160 landslide data verification points and 200 non-landslide data verification points were randomly acquired by manual visual interpretation and field investigation. The validation points were superimposed with the extracted results to determine whether the extracted landslide results are correct or not. It can be obtained through the statistical calculation that the overall accuracy of landslide information extraction from the experimental image 1 by NNC method is 89.5%, with the Kappa coefficient is 0.788, and the overall accuracy by TLOEL method is 90.7%, with Kappa coefficient is 0.812. Table 4 shows the specific results. The overall accuracy of landslide information extraction from the experimental image 2 by NNC method is 90.3%, with the Kappa coefficient is 0.838, and the overall accuracy by the TLOEL method is 91.9%, with the Kappa coefficient is 0.862. Table 5 shows the specific results. The overall accuracy of landslide information extraction from the experimental image 2 by the NNC method is 88.9%, with the Kappa coefficient is 0.842, and the overall accuracy by TLOEL method is 89.4%, with Kappa coefficient is 0.871. Table 6 shows the specific results.

Table 4. Confusion matrix of experimental image 1.

Parameters	NNC Method		TLOEL Method	
	Landslides	Non-landslides	Landslides	Non-landslides
Landslides	239	22	241	17
Non-landslides	41	298	39	303
Producer's precision /%	85.4		86.1	
User's precision /%	91.6		93.4	
Overall precision /%	89.5		90.7	
Kappa coefficient	0.788		0.812	

Table 5. Confusion matrix of experimental image 2.

Parameters	NNC method		TLOEL method	
	Landslides	Non-landslides	Landslides	Non-landslides
Landslides	273	18	287	21
Non-landslides	47	332	33	329
Producer's precision /%	85.3		89.7	
User's precision /%	93.8		92.9	
Overall precision /%	90.3		91.9	
Kappa coefficient	0.838		0.862	

Table 6. Confusion matrix of experimental image 3.

Parameters	NNC method		TLOEL method	
	Landslides	Non-landslides	Landslides	Non-landslides
Landslides	132	12	138	16
Non-landslides	28	188	22	184
Producer's precision/%	82.5		86.2	
User's precision/%	91.7		89.6	
Overall precision/%	88.9		89.4	
Kappa coefficient	0.842		0.871	

5. Discussion

It can be found that the landslides in experimental images 1 and 2 are secondary disasters caused by earthquakes through analysis and field investigation. Most of them are rock landslides, and the spectral and texture characteristics of rock landslides are very similar to those of bare rock masses. This leads to the NNC method misclassifying some bare rock masses into landslides. The TLOEL method has missing classification situation when extracting such landslides. Some of them are muddy landslides; the spectral characteristics of these landslides are similar to those of bare land. NNC method misclassifies some bare land into landslides. Although the TLOEL method has fewer misclassification, it also has some missing classification. The landslide in experiment image 3 is a landslide caused by rainfall. It is difficult to extract the landslide from turbid water and bare rock. The NNC method has the situation of missing classification and misclassifying bare rock into landslides. The TLOEL method has less misclassification, but there are also some missing classifications.

Generally speaking, TLOEL method has certain universality after the completion of sample database construction. For the same task workload, when compared with visual interpretation, this method has obvious advantages in interpretation efficiency, and it has higher accuracy than NNC landslide extraction method. However, in the process of object-oriented image analysis, simple image segmentation will obtain a large number of image objects, and subsequent information extraction will have a large amount of calculation. Landslides usually occur in areas with large topographic fluctuations. In the process of landslide extraction, the Digital Elevation Model (DEM) and slope stability model (e.g., shallow land sliding stability model) can be used to calculate the stability degree of the surface in the certain area [51]. By assigning the stability degree as a weight to the segmented image object, the region with high stability can be eliminated, and the number of objects participating in the subsequent operation can be reduced.

Through the above calculation results, it can be seen that the TLOEL method that is proposed in this paper has advantages for large-scale and scattered distribution landslide extraction, but, at the same time, it is still a complicated work to obtain training samples. In fact, many regions have historical images and historical thematic maps. If these historical data can be used in current interpretation tasks, it will inevitably improve the accuracy and efficiency of interpretation. In future research, besides feature transfer learning, we can also consider design a transfer method for surface feature category labels (associated knowledge) based on two temporal invariant object detection, so as to realize the transfer of "category interpretation knowledge of invariant surface features" from the source domain to the target domain, and to establish a new feature-object mapping relationship.

6. Conclusions

In the paper, wide hazardous areas of landslides in mountainous areas in southwestern China were researched, a landslide sample database was set up, the optimal feature extraction model, namely the transfer learning model, is selected by comparison, and a high-resolution remote sensing image landslide extraction method is proposed by combining this model with the object-oriented image analysis method. This method effectively combines the deep learning in the field of computer with the field of disaster remote sensing, and then improves the automation of landslide information acquisition in the field of high-resolution remote sensing. In addition, the landslide sample database that was established in the paper will provide important data reference for the research of the same type of landslides in southwestern China. While considering that historical archived images and surface feature category maps are available for some researched areas, how to further explore the relationship between historical data and current images and establishing the knowledge transfer framework between historical data and current images will be the key research items in the next step.

Author Contributions: H.L. drafted the manuscript and was responsible for the research design, experiment and analysis. L.M. and N.L. reviewed and edited the manuscript. X.F., Z.W., C.L. and M.T. supported the data preparation and the interpretation of the results. All of the authors contributed to editing and reviewing the manuscript. All authors have read and agreed to the published version of the manuscript.

Remote Sens. **2020**, *12*, 752

Funding: This research was supported by the National Natural Science Foundation of China(41701499), the Sichuan Science and Technology Program(2018GZ0265), the Geomatics Technology and Application Key Laboratory of Qinghai Province, China(QHDX-2018-07), the Major Scientific and Technological Special Program of Sichuan Province, China (2018SZDZX0027), and the Key Research and Development Program of Sichuan Province, China (2018SZ027, 2019-YF09-00081-SN).

Conflicts of Interest: The authors declare no conflict of interest.

References

1. Huang, R.; Fan, X. The landslide story. *Nat. Geosci.* **2013**, *5*, 325–326. [CrossRef]
2. Aksoy, B.; Ercanoglu, M. Landslide identification and classification by object-based image analysis and fuzzy logic: An example from the Azdavay region (Kastamonu, Turkey). *Comput. Geosci.* **2012**, *1*, 87–98. [CrossRef]
3. Pham, B.T.; Shirzadi, A.; Shahabi, H.; Omidvar, E.; Singh, S.K.; Sahana, M.; Asl, D.T.; Ahmad, B.; Quoc, N.K.; Lee, S. Landslide susceptibility assessment by novel hybrid machine learning algorithms. *Sustainability* **2019**, *11*, 4386. [CrossRef]
4. Hong, H.; Chen, W.; Xu, C.; Youssef, A.M.; Pradhan, B.; Tien Bui, D. Rainfall-induced landslide susceptibility assessment at the Chongren area (China) using frequency ratio, certainty factor, and index of entropy. *Geocarto Int.* **2017**, *32*, 139–154. [CrossRef]
5. Pham, B.T.; Prakash, I.; Dou, J.; Singh, S.K.; Trong, P.; Trinh Tran, H.T.; Le, T.M.; Tran, V.P.; Khoi, D.K.; Shirzadi, A.; et al. A novel hybrid approach of landslide susceptibility modeling using rotation forest ensemble and different base classifiers. *Geocarto Int.* **2018**, *10*. [CrossRef]
6. Nguyen, P.T.; Tuyen, T.T.; Shirzadi, A.; Pham, B.T.; Shahabi, H.; Omidvar, E. Development of a novel hybrid intelligence approach for landslide spatial prediction. *Appl. Sci.* **2019**, *9*, 2824. [CrossRef]
7. Dai, F.; Lee, C.; Ngai, Y. Landslide risk assessment and management: An overview. *Eng. Geol.* **2002**, *64*, 65–87. [CrossRef]
8. Pham, B.T.; Prakash, I.; Jaafari, A.; Bui, D.T. Spatial prediction of rainfall-induced landslides using aggregating one-dependence estimators classifier. *J. Indian Soc. Remote Sens.* **2018**, *46*, 1457–1470. [CrossRef]
9. Shafique, M.; van der Meijde, M.; Khan, M.A. A review of the 2005 Kashmir earthquake-induced landslides; from a remote sensing prospective. *J. Asian Earth Sci.* **2016**, *118*, 68–80. [CrossRef]
10. Heleno, S.; Matias, M.; Pina, P.; Sousa, A.J. Semiautomated object-based classification of rain-induced landslides with VHR multispectral images on Madeira island. *Nat. Hazards Earth Syst. Sci.* **2016**, *16*, 1035–1048. [CrossRef]
11. Hölbling, D.; Füreder, P.; Antolini, F.; Cigna, F.; Casagli, N.; Lang, S. A semi-Automated Object-Based Approach for Landslide Detection Validated by Persistent Scatterer Interferometry Measures and Landslide Inventories. *Remote Sens.* **2012**, *4*, 1310–1336. [CrossRef]
12. Sato, H.P.; Hasegawa, H.; Fujiwara, S.; Tobita, M.; Koarai, M.; Une, H.; Iwahashi, J. Interpretation of landslide distribution triggered by the 2005 northern Pakistan earthquake using SPOT 5 imagery. *Landslides* **2007**, *4*, 113–122. [CrossRef]
13. Barlow, J.; Franklin, S.; Martin, Y. High spatial resolution satellite imagery, DEM derivatives, and image segmentation for the detection of mass wasting processes. *Photogramm. Eng. Remote Sens.* **2006**, *72*, 687–692. [CrossRef]
14. Martha, T.R.; Westen, C.J.; Kerle, N.; Jetten, V.; Kumar, K. Landslide hazard and risk assessment using semi-automatically created landslide inventories. *Geomorphology* **2013**, *184*, 139–150. [CrossRef]
15. Van Den Eeckhaut, M.; Kerle, N.; Poesen, J.; Hervás, J. Object-oriented identification of forested landslides with derivatives of single pulse LiDAR data. *Geomorphology* **2012**, *173*, 30–42. [CrossRef]
16. Blaschke, T.; Feizizadeh, B.; Hölbling, D. Object-based image analysis and digital terrain analysis for locating landslides in the Urmia Lake Basin, Iran. *IEEE J. Sel. Top. Appl. Earth Obs. Remote Sens.* **2014**, *7*, 4806–4817. [CrossRef]
17. Guirado, E.; Tabik, S.; Alcaraz-Segura, D.; Cabello, J.; Herrera, F. Deep-learning versus OBIA for scattered shrub detection with Google Earth imagery: Ziziphus Lotus as case study. *Remote Sens.* **2017**, *9*, 1220. [CrossRef]
18. Domingos, P. A few useful things to know about machine learning. *Commun. ACM* **2012**, *55*, 78–87. [CrossRef]
19. Cheng, G.; Han, J. A survey on object detection in optical remote sensing images. *ISPRS J. Photogramm. Remote Sens.* **2016**, *117*, 11–28. [CrossRef]

20. Chen, W.; Xie, X.; Wang, J.; Pradhan, B.; Hong, H.; Bui, D.T.; Duan, Z.; Ma, J. A comparative study of logistic model tree, random forest, and classification and regression tree models for spatial prediction of landslide susceptibility. *Catena* **2017**, *151*, 147–160. [CrossRef]

21. Längkvist, M.; Kiselev, A.; Alirezaie, M.; Loutfi, A. Classification and segmentation of satellite orthoimagery using convolutional neural networks. *Remote Sens.* **2016**, *8*, 329. [CrossRef]

22. Maggiori, E.; Tarabalka, Y.; Charpiat, G.; Alliez, P. Convolutional neural networks for large-scale remote-sensing image classification. *IEEE Trans. Geosci. Remote Sens.* **2017**, *55*, 645–657. [CrossRef]

23. Liu, T.; Elrahman, A. Comparing fully convolutional networks, random forest, support vector machine, and patch-based deep convolutional neural networks for object-based wetland mapping using images from small unmanned aircraft system. *GISci. Remote Sens.* **2018**, *55*, 243–264. [CrossRef]

24. Hinton, G.E.; Salakhutdinov, R.R. Reducing the dimensionality of data with neural networks. *Science* **2006**, *313*, 504–507. [CrossRef] [PubMed]

25. Lee, H. Convolutional deep belief networks for scalable unsupervised learning of hierarchical representations. In Proceedings of the 26th Annual International Conference on Machine Learning, Montreal, QC, Canada, 14–18 June 2009.

26. Jia, Y. Caffe: Convolutional architecture for fast feature embedding. In Proceedings of the ACM International Conference on Multimedia, Orlando, FL, USA, 3–7 November 2014.

27. Laliberte, A.S.; Rango, A. Image processing and classification procedures for analysis of sub-decimeter imagery acquired with an unmanned aircraft over arid rangelands. *GISci. Remote Sens.* **2011**, *48*, 4–23. [CrossRef]

28. Rango, A.; Laliberte, A.S. Impact of flight regulations on effective use of unmanned aircraft systems for natural resources applications. *J. Appl. Remote Sens.* **2010**, *4*, 043539.

29. Trimble GmbH. *eCognition Developer 9.0 User Guide*; Trimble Germany GmbH: Munich, Germany, 2014.

30. He, K. Deep residual learning for image recognition. In Proceedings of the IEEE Conference on Computer Vision and Pattern Recognition, Las Vegas, NV, USA, 26 June–1 July 2016; pp. 770–778.

31. Dalal, N.; Triggs, B. Histograms of oriented gradients for human detection. In Proceedings of the 2005 IEEE Computer Society Conference on Computer Vision and Pattern Recognition, San Diego, CA, USA, 20–26 June 2005; Volume 1, pp. 886–893.

32. Lu, P.; Bai, S.; Casagli, N. Investigating spatial patterns of persistent scatter interferometry point targets and landslide occurrences in the Arno river basin. *Remote Sens.* **2014**, *6*, 6817–6843. [CrossRef]

33. Tarolli, P.; Sofia, G.; Dalla fontana, G. Geomorphic features extraction from high-resolution topography: Landslide crowns and bank erosion. *Nat. Hazards* **2012**, *61*, 65–83. [CrossRef]

34. Lacroix, P.; Zavala, B.; Berthier, E. Supervised method of landslide inventory using panchromatic spot5 images and application to the earthquake-triggered landslides of Pisco (peru, 2007, Mw8.0). *Remote Sens.* **2013**, *5*, 2590–2616. [CrossRef]

35. Wiegand, C.; Rutzinger, M.; Heinrich, K. Automated extraction of shallow erosion areas based on multi-temporal ortho-imagery. *Remote Sens.* **2013**, *5*, 2292–2307. [CrossRef]

36. Cheng, G.; Guo, L.; Zhao, T. Automatic landslide detection from remote-sensing imagery using a scene classification method based on Bovw and Plsa. *Int. J. Remote Sens.* **2013**, *34*, 45–59. [CrossRef]

37. Lu, P.; Stumpf, A.; Kerle, N. Object-oriented change detection for landslide rapid mapping. *IEEE Geosci. Remote Sens. Lett.* **2011**, *8*, 701–705. [CrossRef]

38. Stumpf, A.; Lachiche, N.; Malet, J. Active learning in the spatial domain for remote sensing image classification. *IEEE Trans. Geosci. Remote Sens.* **2014**, *52*, 2492–2507. [CrossRef]

39. Debella-gilo, m.; Kääb, A. Measurement of Surface Displacement and Deformation of Mass Movements Using Least Squares Matching of Repeat High Resolution Satellite and Aerial Images. *Remote Sens.* **2012**, *4*, 43–67. [CrossRef]

40. Barazzetti, L.; Scaioni, M.; Gianinetto, M. Automatic co-registration of satellite time series via least squares adjustment. *Eur. J. Remote Sens.* **2014**, *47*, 55–74. [CrossRef]

41. Liao, M.; Tang, J.; Wang, T. Landslide monitoring with high-resolution SAR data in the three Gorges region. *Sci. China Earth Sci.* **2012**, *55*, 590–601. [CrossRef]

42. Ventisette, C.D.; Intrieri, E.; Luzi, G. Using ground based radar interferometry during emergency: The case of the A3 Motorway (calabria Region, Italy) threatened By a landslide. *Nat. Hazards Earth Syst. Sci.* **2011**, *11*, 2483–2495. [CrossRef]

43. Bai, S.; Wang, J.; Lv, G.N. GIS-based logistic regression for landslide susceptibility mapping of the Zhongxian segment in the three Gorges Area, China. *Geomorphology* **2009**, *115*, 23–31. [CrossRef]
44. Xu, C.; Xu, X.W.; Dai, F.C. Application of an incomplete landslide inventory, logistic regression model and its validation for landslide susceptibility mapping related to the 12 May 2008 Wenchuan earthquake of China. *Nat. Hazards* **2013**, *68*, 883–900. [CrossRef]
45. Bai, S.; Wang, J.; Zhang, Z. Combined landslide susceptibility mapping after Wenchuan earthquake at the Zhouqu segment in the Bailongjiang basin, China. *Catena* **2012**, *99*, 18–25. [CrossRef]
46. Fiorucci, F.; Cardinali, M.; Carlà, R. Seasonal landslide mapping and estimation of landslide mobilization rates using aerial and satellite images. *Geomorphology* **2011**, *129*, 59–70. [CrossRef]
47. Galarreta, J.F.; Kerle, N.; Gerke, M. UAV-based urban structural damage assessment using object-based image analysis and semantic reasoning. *Nat. Hazards Earth Syst. Sci.* **2015**, *15*, 1087. [CrossRef]
48. Duro, D.C.; Franklin, S.E.; Dubé, M.G. Multi-scale object-based image analysis and feature selection of multi-sensor earth observation imagery using random forests. *Int. J. Remote Sens.* **2012**, *33*, 4502–4526. [CrossRef]
49. Foody, G.M. Status of land cover classification accuracy assessment. *Remote Sens. Environ.* **2002**, *80*, 185–201. [CrossRef]
50. Dragut, L.; Tiede, D.; Shaun, R. ESP: A tool to estimate scale parameter for multiresolution image segmentation of remotely sensed data. *Int. J. Geogr. Inf. Sci.* **2010**, *6*, 859–871. [CrossRef]
51. Montgomery, D.R.; Dietrich, W.E. Channel Initiation and the Problem of Landscape Scale. *Science* **1994**, *255*, 826–830. [CrossRef]

MDPI

St. Alban-Anlage 66

4052 Basel

Switzerland

Tel. +41 61 683 77 34

Fax +41 61 302 89 18

www.mdpi.com

Remote Sensing Editorial Office

E-mail: remotesensing@mdpi.com

www.mdpi.com/journal/remotesensing